Cram101 Textbook Outlines to accompany:

Evolutionary Analysis

Freeman, Scott, 4th Edition

With Cram101.com online, you also have access to extensive reference material.

You will nail those essays and papers. Here is an example from a Cram101 Biology text:

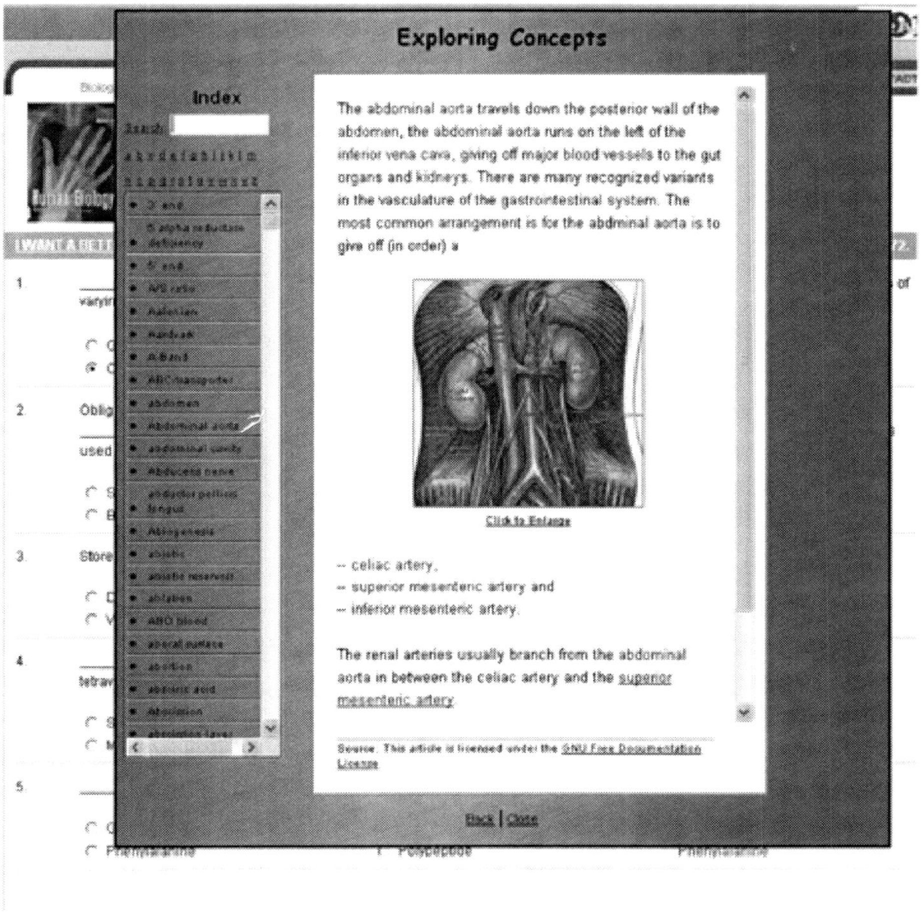

Visit **www.Cram101.com**, click Sign Up at the top of the screen, and enter DK73DW5994 in the promo code box on the registration screen. Access to www.Cram101.com is normally $9.95 per month, but because you have purchased this book, your access fee is only $4.95 per month, cancel at any time. Sign up and stop highlighting textbooks forever.

Learning System

Cram101 Textbook Outlines is a learning system. The notes in this book are the highlights of your textbook, you will never have to highlight a book again.

How to use this book. Take this book to class, it is your notebook for the lecture. The notes and highlights on the left hand side of the pages follow the outline and order of the textbook. All you have to do is follow along while your instructor presents the lecture. Circle the items emphasized in class and add other important information on the right side. With Cram101 Textbook Outlines you'll spend less time writing and more time listening. Learning becomes more efficient.

Cram101.com Online

Increase your studying efficiency by using Cram101.com's practice tests and online reference material. It is the perfect complement to Cram101 Textbook Outlines. Use self-teaching matching tests or simulate in-class testing with comprehensive multiple choice tests, or simply use Cram's true and false tests for quick review. Cram101.com even allows you to enter your in-class notes for an integrated studying format combining the textbook notes with your class notes.

Visit **www.Cram101.com**, click Sign Up at the top of the screen, and enter **DK73DW5994** in the promo code box on the registration screen. Access to www.Cram101.com is normally $9.95 per month, but because you have purchased this book, your access fee is only $4.95 per month. Sign up and stop highlighting textbooks forever.

 ISBN(s): 9781616986278. EDR-5.2010111

Evolutionary Analysis
Freeman, Scott, 4th

CONTENTS

AIDS

Acquired immune deficiency syndrome or acquired immunodeficiency syndrome (AIDS) is a disease of the human immune system caused by the human immunodeficiency virus (HIV.)

This condition progressively reduces the effectiveness of the immune system and leaves individuals susceptible to opportunistic infections and tumors. HIV is transmitted through direct contact of a mucous membrane or the bloodstream with a bodily fluid containing HIV, such as blood, semen, vaginal fluid, preseminal fluid, and breast milk.

Charles Robert Darwin

Charles Robert Darwin FRS (12 February 1809 - 19 April 1882) was an English naturalist who realised and presented compelling evidence that all species of life have evolved over time from common ancestors, through the process he called natural selection. The fact that evolution occurs became accepted by the scientific community and much of the general public in his lifetime, while his theory of natural selection came to be widely seen as the primary explanation of the process of evolution in the 1930s, and now forms the basis of modern evolutionary theory. In modified form, Darwin"s scientific discovery is the unifying theory of the life sciences, providing logical explanation for the diversity of life.

Human

A Human is a member of a species of bipedal primates in the family Hominidae . DNA and fossil evidence indicates that modern Human s originated in east Africa about 200,000 years ago. When compared to other animals and primates, Human s have a highly developed brain, capable of abstract reasoning, language, introspection and problem solving.

Species

In biology, a Species is:

· a taxonomic rank (the basic rank of biological classification) or
· a unit at that rank

There are many definitions of what kind of unit a Species is (or should be.) A common definition is that of a group of organisms capable of interbreeding and producing fertile offspring, and separated from other such groups with which interbreeding does not (normally) happen. Other definitions may focus on similarity of DNA or morphology. Some Species are further subdivided into sub Species , and here also there is no close agreement on the criteria to be used.

Virus

A virus is a microscopic infectious agent that can reproduce only inside a host cell. virus es infect all types of organisms: from animals and plants, to bacteria and archaea. Since the initial discovery of tobacco mosaic virus by Martinus Beijerinck in 1898, more than 5,000 types of virus have been described in detail, although most types of virus remain undiscovered.

Fibroblast

A Fibroblast is a type of cell that synthesizes the extracellular matrix and collagen, the structural framework (stroma) for animal tissues, and play a critical role in wound healing. They are the most common cells of connective tissue in animals.

Fibroblast s and fibrocytes are two states of the same cells, the former being the activated state, the latter the less active state, concerned with maintenance.

Fibroblast growth factor	Fibroblast growth factor s are a family of growth factors involved in angiogenesis, wound healing, and embryonic development. The Fibroblast growth factor s are heparin-binding proteins and interactions with cell-surface associated heparan sulfate proteoglycans have been shown to be essential for Fibroblast growth factor signal transduction. Fibroblast growth factor s are key players in the processes of proliferation and differentiation of wide variety of cells and tissues.
Plague	Plague is a deadly infectious disease caused by the enterobacteria Yersinia pestis (Pasteurella pestis.) Plague is a zoonotic, primarily carried by rodents (most notably rats) and spread to humans via fleas. Plague is notorious throughout history, due to the unprecedented scale of death and devastation it brought.
Protein	Protein s are organic compounds made of amino acids arranged in a linear chain. The amino acids in a polymer chain are joined together by the peptide bonds between the carboxyl and amino groups of adjacent amino acid residues. The sequence of amino acids in a protein is defined by the sequence of a gene, which is encoded in the genetic code.
Rat	Rat s are various medium-sized, long-tailed rodents of the superfamily Muroidea. "True Rat s" are members of the genus Rat tus, the most important of which to humans are the black Rat Rat tus Rat tus, and the brown Rat Rat tus norvegicus. Many members of other rodent genera and families are also called Rat s and share many characteristics with true Rat s.
Adaptation	In ocular physiology, Adaptation is the ability of the eye to adjust to various levels of darkness and light. The human eye can function from very dark to very bright levels of light -- its sensing capabilities reach across nine orders of magnitude. This means that the brightest and the darkest light signal that the eye can sense are a factor of roughly one thousand million apart.
Leukemia inhibitory factor	Leukemia inhibitory factor an interleukin 6 class cytokine, is a chemical in cells that affects their growth and development. Leukemia inhibitory factor derives its name from its ability to induce the terminal differentiation of myeloid leukaemic cells. Other properties attributed to the cytokine include: the growth promotion and cell differentiation of different types of target cells, influence on bone metabolism, cachexia, neural development, embryogenesis and inflammation.
Life expectancy	Life expectancy is the average number of years of life remaining at a given age. The term is most often used in the human context, but used also in plant or animal ecology and the calculation is based on the analysis of life tables (also known as actuarial tables.) The term may also be used in the context of manufactured objects although the related term shelf life is used for consumer products.

Reverse transcriptase

In biochemistry, a Reverse transcriptase is a DNA polymerase enzyme that transcribes single-stranded RNA into single-stranded DNA. It also helps in the formation of a double helix DNA once the RNA has been reverse transcribed into a single strand cDNA. Normal transcription involves the synthesis of RNA from DNA; hence, reverse transcription is the reverse of this.

Well studied Reverse transcriptase s include:

· HIV-1 Reverse transcriptase from human immunodeficiency virus type 1
· M-MLV Reverse transcriptase from the Moloney murine leukemia virus
· AMV Reverse transcriptase from the avian myeloblastosis virus
· Telomerase Reverse transcriptase that maintains the telomeres of eukaryotic chromosomes

The enzyme is encoded and used by reverse-transcribing viruses, which use the enzyme during the process of replication. Reverse-transcribing RNA viruses, such as retroviruses, use the enzyme to reverse-transcribe their RNA genomes into DNA, which is then integrated into the host genome and replicated along with it.

Beak

The Beak bill or rostrum is an external anatomical structure of birds which is used for eating and for grooming, manipulating objects, killing prey, probing for food, courtship and feeding young. The term also refers to a similar mouthpart in some monotremes, cephalopods, cetaceans, pufferfishes, turtles, Anuran tadpoles and sirens. Comparison of bird Beak s, displaying different shapes adapted to different feeding methods.

Cell

The Cell is the structural and functional unit of all known living organisms. It is the smallest unit of an organism that is classified as living, and is often called the building block of life. Some organisms, such as most bacteria, are unicellular (consist of a single Cell.)

Effector

An Effector is a molecule (originally referring to small molecules but now encompassing any regulatory molecule, including proteins) that binds to a protein and thereby alters the activity of that protein. A modulator molecule binds to a regulatory site during allosteric modulation and allosterically modulates the shape of the protein.

· Activators
· Inhibitors. .

Simian immunodeficiency viruses

Simian immunodeficiency viruses are retroviruses that are found in over 40 African primates.

Viruses from two of these primate species, Simian immunodeficiency virusessmm in Sooty Mangabeys and Simian immunodeficiency virusescpz in chimpanzees, are believed to have crossed the species barrier into humans, resulting in HIV-2 and HIV-1, respectively. The most likely route of transmission of HIV-1 to humans involves contact with the blood of chimps that are often hunted for bushmeat in Africa.

Macrophages	Macrophages are white blood cells within tissues, produced by the division of monocytes. Human Macrophages are about 21 micrometres in diameter. Monocytes and Macrophages are phagocytes, acting in both non-specific defense as well as to help initiate specific defense mechanisms (or adaptive immunity) of vertebrate animals.
Thymidine	Thymidine (more precisely called deoxy Thymidine ; can also be labelled deoxyribosylthymine, and thymine deoxyriboside) is a chemical compound, more precisely a pyrimidine deoxynucleoside. Deoxy Thymidine is the DNA nucleoside T, which pairs with deoxyadenosine (A) in double-stranded DNA. In cell biology it is used to synchronize the cells in S phase. Before the boom in Thymidine use caused by the need for Thymidine for the production of the antiretroviral drug azido Thymidine (AZT), much of the world"s Thymidine production came from herring sperm.
Mutation	In biology, mutation s are changes to the nucleotide sequence of the genetic material of an organism. mutation s can be caused by copying errors in the genetic material during cell division, by exposure to ultraviolet or ionizing radiation, chemical mutagens, or viruses, or can be induced by the organism itself, by cellular processes such as hyper mutation . In multicellular organisms with dedicated reproductive cells, mutation s can be subdivided into germ line mutation s, which can be passed on to descendants through the reproductive cells, and somatic mutation s, which involve cells outside the dedicated reproductive group and which are not usually transmitted to descendants.
Back	The Back describes the area of horse anatomy where the saddle goes, and in popular usage extends to include the loin or lumbar region behind the thoracic vertebrae that also is crucial to a horse"s weight-carrying ability. These two sections of the vertebral column beginning at the withers, the start of the thoracic vertebrae, and extend to the last lumbar vertebra. Because horses are ridden by humans, the strength and structure of the horse"s Back is critical to the animal"s usefulness.
Combination	In combinatorial mathematics, a combination is an un-ordered collection of distinct elements, usually of a prescribed size and taken from a given set. (An ordered collection of distinct elements would sometimes be called a permutation, but that term is ambiguous; it can also mean "reordering of all terms", among other related notions.) Given such a set S, a combination of elements of S is just a subset of S, where, as always for (sub)sets the order of the elements is not taken into account (two lists with the same elements in different orders are considered to be the same combination.)
Entry inhibitors	Entry inhibitors are a class of antiretroviral drugs, used in combination therapy for the treatment of HIV infection. This class of drugs interferes with the binding, fusion and entry of an HIV virion to a human cell. By blocking this step in HIV"s replication cycle, such agents slow the progression from HIV infection to AIDS. An a HIV virion binds to a CD4+ human cell.

Genetic	Genetics is the study of how living things receive common traits from previous generations. These traits are described by the Genetic information carried by a molecule called DNA. The instructions for constructing and operating an organism are contained in the organism"s DNA. Every living thing on earth has DNA in its cells. Genes are the hereditary components of DNA that occupy spots on chromosomes and determine characteristics in an organism.
Genetic variation	Genetic variation, variation in alleles of genes, occurs both within and among populations. genetic variation is important because it provides the "raw material" for natural selection. genetic variation among individuals within a population can be identified at a variety of levels.
Integrase inhibitor	Integrase inhibitor s are a class of antiretroviral drug designed to block the action of integrase, a viral enzyme that inserts the viral genome into the DNA of the host cell. Since integration is a vital step in retroviral replication, blocking it can halt further spread of the virus. Integrase inhibitor s were initially developed for the treatment of HIV infection, but they could be applied to other retroviruses.
Protease	A Protease is any enzyme that conducts proteolysis, that is, begins protein catabolism by hydrolysis of the peptide bonds that link amino acids together in the polypeptide chain, which form a molecule of protein. In other words, Protease s break down proteins. Protease s are currently classified into six groups: · Serine Protease s · Threonine Protease s · Cysteine Protease s · Aspartate Protease s · Metallo Protease s · Glutamic acid Protease s The threonine and glutamic acid Protease s were not described until 1995 and 2004, respectively. The mechanism used to cleave a peptide bond involves making an amino acid residue that has the cysteine and threonine (peptidases) or a water molecule (aspartic acid, metallo- and glutamic acid peptidases) nucleophilic so that it can attack the peptide carboxyl group.
Reverse transcriptase inhibitors	Reverse transcriptase inhibitors (Reverse transcriptase inhibitors s) are a class of antiretroviral drug used to treat HIV infection, tumors, and cancer. Reverse transcriptase inhibitors s inhibit activity of reverse transcriptase, a viral DNA polymerase enzyme that retroviruses need to reproduce. When HIV infects a cell, reverse transcriptase copies the viral single stranded RNA genome into a double-stranded viral DNA. The viral DNA is then integrated into the host chromosomal DNA, which then allows host cellular processes, such as transcription and translation to reproduce the virus.

Genome	In classical genetics, the Genome of a diploid organism including eukarya refers to a full set of chromosomes or genes in a gamete; thereby, a regular somatic cell contains two full sets of Genome s. In haploid organisms, including bacteria, archaea, viruses, and mitochondria, a cell contains only a single set of the Genome usually in a single circular or contiguous linear DNA (or RNA for retroviruses.) In modern molecular biology the Genome of an organism is its hereditary information encoded in DNA (or, for retroviruses, RNA.)
Chromosome	A Chromosome is an organized structure of DNA and protein that is found in cells. It is a single piece of coiled DNA containing many genes, regulatory elements and other nucleotide sequences. Chromosome s also contain DNA-bound proteins, which serve to package the DNA and control its functions.
Allele	An Allele) is one of a series of different forms of a gene. The word is a short from of allelomorph , which was used in the early days of genetics to describe variant forms of a gene detected as different phenotypes. Allele s are now understood to be alternative DNA sequences at the same physical gene locus, which may or may not result in different phenotypic traits.
Chicken anaemia virus	Chicken anaemia virus is a virus that affects poultry. Chicken anaemia virus causes anaemia, bone marrow atrophy, and severe immunosuppression. Clinical signs of infection of Chicken anaemia virus is predominantly found in young chicks due to maternal antibodies present in most adult chickens.
Edward	The Edward mango is a monoembryonic mango cultivar grown predominantly in Florida. It is considered by many to be among the finest tasting mangoes in the world; however, its poor yields have restrained the Edward from developing into a commercially significant variety. The Edward was first propagated in the 1920s by Edward Simmonds of the Plant Introduction Garden in Miami, Florida and is believed to be a hybrid cross of Haden and Carabao mango cultivars.
Genetic drift	Genetic drift or allelic drift is the change in the relative frequency with which a gene variant (allele) occurs in a population that results from the fact that alleles in offspring are a random sample of those in the parents, and because of the role of chance in determining whether a given individual survives and reproduces. A population"s allele frequency is the fraction of the gene copies that share a particular form. Genetic drift is one of several evolutionary processes which lead to changes in allele frequencies over time.
Chimpanzee	Chimpanzee sometimes colloquially chimp, is the common name for the two extant species of ape in the genus Pan where the Congo River forms the boundary between the native habitat of the two species:

	· Common Chimpanzee Pan troglodytes: the better known Chimpanzee lives primarily in West and Central Africa. · Bonobo, Pan paniscus: also known as the "Pygmy Chimpanzee , this species is found in the forests of the Democratic Republic of the Congo. Chimpanzee s are members of the Hominidae family, along with gorillas, humans, and orangutans. Chimpanzee are thought to have split from human evolution about 6 million years ago and thus the two Chimpanzee species are the closest living relatives to humans, all being members of the Hominini tribe (along with extinct species of Hominina subtribe.) Chimpanzee s are the only known members of the Panina subtribe. The two Pan species split only about one million years ago.
Vaccine	A vaccine is a biological preparation that improves immunity to a particular disease. A vaccine typically contains a small amount of an agent that resembles a microorganism. The agent stimulates the body"s immune system to recognize the agent as foreign, destroy it, and "remember" it, so that the immune system can more easily recognize and destroy any of these microorganisms that it later encounters.
Key Habitat Site	A Key Habitat Site is a Canadian Wildlife Service designation for an area that supports at least 1% of the country"s population of any migratory bird species at any time. There may be overlap with areas designated as a migratory bird sanctuary or national wildlife area.
Blytt-Sernander	The Blytt-Sernander classification is a series of north European climatic periods or phases based on the study of Danish peat bogs by Axel Blytt (1876) and Rutger Sernander (1908.) The classification was incorporated into a sequence of pollen zones later defined by Lennart von Post, one of the founders of palynology. Layers in peat were first noticed by Heinrich Dau in 1829.

Snake	Snakes, like other reptiles, have a skin covered in scales. Snakes are entirely covered with scales or scutes of various shapes and sizes. Scales protect the body of the Snake, aid it in locomotion, allow moisture to be retained within, alter the surface characteristics such as roughness to aid in camouflage, and in some cases even aid in prey capture (such as Acrochordus.)
Charles Robert Darwin	Charles Robert Darwin FRS (12 February 1809 - 19 April 1882) was an English naturalist who realised and presented compelling evidence that all species of life have evolved over time from common ancestors, through the process he called natural selection. The fact that evolution occurs became accepted by the scientific community and much of the general public in his lifetime, while his theory of natural selection came to be widely seen as the primary explanation of the process of evolution in the 1930s, and now forms the basis of modern evolutionary theory. In modified form, Darwin"s scientific discovery is the unifying theory of the life sciences, providing logical explanation for the diversity of life.
Species	In biology, a Species is: · a taxonomic rank (the basic rank of biological classification) or · a unit at that rank There are many definitions of what kind of unit a Species is (or should be.) A common definition is that of a group of organisms capable of interbreeding and producing fertile offspring, and separated from other such groups with which interbreeding does not (normally) happen. Other definitions may focus on similarity of DNA or morphology. Some Species are further subdivided into sub Species , and here also there is no close agreement on the criteria to be used.
Adaptation	In ocular physiology, Adaptation is the ability of the eye to adjust to various levels of darkness and light. The human eye can function from very dark to very bright levels of light -- its sensing capabilities reach across nine orders of magnitude. This means that the brightest and the darkest light signal that the eye can sense are a factor of roughly one thousand million apart.
Genetic	Genetics is the study of how living things receive common traits from previous generations. These traits are described by the Genetic information carried by a molecule called DNA. The instructions for constructing and operating an organism are contained in the organism"s DNA. Every living thing on earth has DNA in its cells. Genes are the hereditary components of DNA that occupy spots on chromosomes and determine characteristics in an organism.
Population	In biology, a population is the collection of inter-breeding organisms of a particular species; in sociology, a collection of human beings. Individuals within a population share a factor may be reduced by statistical means, but such a generalization may be too vague to imply anything. Demography is used extensively in marketing, which relates to economic units, such as retailers, to potential customers.

Scale	In most biological nomenclature, a Scale is a small rigid plate that grows out of an animal"s skin to provide protection. In lepidopteran species, scales are plates on the surface of the insect wing, and provide coloration. Scales are quite common and have evolved multiple times with varying structure and function.
Trait	A trait is a distinct variant of a phenotypic character of an organism that may be inherited, environmentally determined or somewhere in between. For example, eye color is a character or abstraction of an attribute, while blue, brown and hazel are traits. A trait may be any single feature or quantifiable measurement of an organism.
Armadillo	Armadillo s are small placental mammals, known for having a leathery armor shell. The Dasypodidae are the only surviving family in the order Cingulata, part of the superorder Xenarthra along with the anteaters and sloths. The word Armadillo is Spanish for "little armored one".
Feathers	Feathers are one of the epidermal growths that form the distinctive outer covering on birds. They are considered the most complex integumentary structures found in vertebrates. They are among the outstanding characteristics that distinguish the extant Aves from other living groups.
Dinosaur	Dinosaur s were the dominant vertebrate animals of terrestrial ecosystems for over 160 million years, from the late Triassic period (about 230 million years ago) until the end of the Cretaceous period (65 million years ago), when most of them became extinct in the Cretaceous-Tertiary extinction event. The 10 000 living species of birds may be classified as Dinosaur s. The term Dinosaur was coined in 1842 by Sir Richard Owen and derives from Greek δεινΪŒς "terrible, powerful, wondrous" + σαά¿¦ρος (sauros) "lizard".
Form	In botanical nomenclature, a Form is a low-level taxonomic rank below that of variety; it is an infraspecific taxon. Its name consists of three parts: a genus name, a specific epithet, and an infraspecific epithet. The abbreviation "f." or the full "forma" should be put before the infraspecific epithet to indicate the rank.
Isis	Isis is a genus of deep-sea bamboo coral of the Isididae family.
Branch	A Branch or tree Branch is a woody structural member connected to but not part of the central trunk of a tree (or sometimes a shrub.) Large Branch es are known as boughs and small Branch es are known as twigs. While Branch es can be nearly horizontal, vertical, or diagonal, the majority of trees have upwardly diagonal Branch es.
Snail	The word Snail is a common name for almost all members of the molluscan class Gastropoda that have coiled shells in the adult stage. When the word Snail is used in a general sense, it includes sea Snail s, land Snail s and freshwater Snail s. Snail s lacking a shell or having only a very small one are usually called slugs.

Leukemia inhibitory factor

Leukemia inhibitory factor an interleukin 6 class cytokine, is a chemical in cells that affects their growth and development.

Leukemia inhibitory factor derives its name from its ability to induce the terminal differentiation of myeloid leukaemic cells. Other properties attributed to the cytokine include: the growth promotion and cell differentiation of different types of target cells, influence on bone metabolism, cachexia, neural development, embryogenesis and inflammation.

Chloramphenicol acetyltransferase

Chloramphenicol acetyltransferase (or Chloramphenicol acetyltransferase T) is a bacterial enzyme that detoxifies the antibiotic chloramphenicol and is responsible for chloramphenicol resistance in bacteria. This enzyme covalently attaches an acetyl group from acetyl-CoA to chloramphenicol, which prevents chloramphenicol from binding to ribosomes.

Chloramphenicol acetyltransferase T is used as a reporter system to measure the level of a promoter or its tissue-specific expression.

Treatise on Invertebrate Paleontology

The Treatise on Invertebrate Paleontology published by the Geological Society of America and the University of Kansas Press, is a definitive multi-authored work of some 50 volumes, written by more than 300 paleontologists, and covering every phylum, class, order, family, and genus of fossil and extant (still living) invertebrate animals. The prehistoric invertebrates are described as to their taxonomy, morphology, paleoecology, stratigraphic and paleogeographic range. However, genera with no fossil record whatsoever have just a very brief listing.

Forelimb

A Forelimb is an anterior (front-most) limb (arm, leg, or similar appendage) on an animal"s body. When referring to quadrupeds (four-legged animals), the term foreleg is often instead used.

The term is not to be confused with the forearm, which refers to a segment of the arm between the elbow and the wrist.

Irwin

The Irwin mango is a commercial mango cultivar which was developed in South Florida.

The Irwin was a seedling of the Lippens cultivar that was open-cross pollinated with Haden, planted on the property of F.D. Irwin in Miami, Florida in 1939. The tree first bore fruit in 1945 and was named and described in 1949.

Vertebrates

Vertebrates are members of the subphylum Vertebrata, chordates with backbones or spinal columns. The grouping sometimes includes the hagfish, which have no vertebrae, but are genetically quite closely related to lampreys, which do have vertebrae. For this reason, the sub-phylum is sometimes referred to as "Craniata", as all members do possess a cranium.

Embryo

An Embryo is a multicellular diploid eukaryote in its earliest stage of development, from the time of first cell division until birth, hatching, or germination. In humans, it is called an Embryo until about eight weeks after fertilization (i.e. ten weeks LMP), and from then it is instead called a fetus. 6 week old excised human Embryo /span>

The development of the Embryo is called Embryo genesis.

Flower

A Flower sometimes known as a bloom or blossom, is the reproductive structure found in Flower ing plants The biological function of a Flower is to mediate the union of male sperm with female ovum in order to produce seeds. The process begins with pollination, is followed by fertilization, leading to the formation and dispersal of the seeds.

Chromosome

A Chromosome is an organized structure of DNA and protein that is found in cells. It is a single piece of coiled DNA containing many genes, regulatory elements and other nucleotide sequences. Chromosome s also contain DNA-bound proteins, which serve to package the DNA and control its functions.

Genetic code

The Genetic code is the set of rules by which information encoded in genetic material (DNA or RNA sequences) is translated into proteins (amino acid sequences) by living cells. The code defines a mapping between tri-nucleotide sequences, called codons, and amino acids. A triplet codon in a nucleic acid sequence usually specifies a single amino acid (though in some cases the same codon triplet in different locations can code unambiguously for two different amino acids, the correct choice at each location being determined by context.)

Acid

An acid is traditionally considered any chemical compound that, when dissolved in water, gives a solution with a hydrogen ion activity greater than in pure water, i.e. a pH less than 7.0. That approximates the modern definition of Johannes Nicolaus Brønsted and Martin Lowry, who independently defined an acid as a compound which donates a hydrogen ion (H^+) to another compound (called a base.) Common examples include acetic acid and sulfuric acid (used in car batteries.)

Protein

Protein s are organic compounds made of amino acids arranged in a linear chain. The amino acids in a polymer chain are joined together by the peptide bonds between the carboxyl and amino groups of adjacent amino acid residues. The sequence of amino acids in a protein is defined by the sequence of a gene, which is encoded in the genetic code.

Charcot-Marie-Tooth disease

Charcot-Marie-Tooth disease, known also as Hereditary Motor and Sensory Neuropathy (HMSN), Hereditary Sensorimotor Neuropathy (HSMN) is a heterogeneous inherited disorder of nerves (neuropathy) that is characterized by loss of muscle tissue and touch sensation, predominantly in the feet and legs but also in the hands and arms in the advanced stages of disease. Presently incurable, this disease is one of the most common inherited neurological disorders, with 37 in 100,000 affected.

The disorder is caused by the absence of proteins that are essential for normal function of the nerves due to errors in the gene coding molecules.

Intron

An Intron is a DNA region within a gene that is not translated into protein. These non-coding sections are transcribed to precursor mRNA (pre-mRNA) and some other RNAs (such as long noncoding RNAs), and subsequently removed by a process called splicing during the processing to mature RNA. After Intron splicing (ie. removal), the mRNA consists only of exon derived sequences, which are translated into a protein.

Retrotransposons

Retrotransposons are genetic elements that can amplify themselves in a genome and are ubiquitous components of the DNA of many eukaryotic organisms. They are a subclass of transposon. They are particularly abundant in plants, where they are often a principal component of nuclear DNA. In maize, 49-78% of the genome is made up of Retrotransposons.

Reverse transcriptase

In biochemistry, a Reverse transcriptase is a DNA polymerase enzyme that transcribes single-stranded RNA into single-stranded DNA. It also helps in the formation of a double helix DNA once the RNA has been reverse transcribed into a single strand cDNA. Normal transcription involves the synthesis of RNA from DNA; hence, reverse transcription is the reverse of this.

Well studied Reverse transcriptase s include:

· HIV-1 Reverse transcriptase from human immunodeficiency virus type 1
· M-MLV Reverse transcriptase from the Moloney murine leukemia virus
· AMV Reverse transcriptase from the avian myeloblastosis virus
· Telomerase Reverse transcriptase that maintains the telomeres of eukaryotic chromosomes

The enzyme is encoded and used by reverse-transcribing viruses, which use the enzyme during the process of replication. Reverse-transcribing RNA viruses, such as retroviruses, use the enzyme to reverse-transcribe their RNA genomes into DNA, which is then integrated into the host genome and replicated along with it.

Escherichia

Escherichia is a genus of Gram-negative, non-spore forming, facultatively anaerobic, rod-shaped bacteria from the family Enterobacteriaceae. Inhabitants of the gastrointestinal tracts of warm-blooded animals, Escherichia species provide a portion of the microbially-derived vitamin K for their host.

While many Escherichia are harmless commensals, particular strains of some species are human pathogens, and are known as the most common cause of urinary tract infections, significant sources of gastrointestinal disease, ranging from simple diarrhea to dysentery-like conditions, as well as a wide-range of other pathogenic states.

Escherichia coli

Escherichia coli , is a Gram negative bacterium that is commonly found in the lower intestine of warm-blooded organisms . Most E. coli strains are harmless, but some, such as serotype O157:H7, can cause serious food poisoning in humans, and are occasionally responsible for costly product recalls. The harmless strains are part of the normal flora of the gut, and can benefit their hosts by producing vitamin K_2, or by preventing the establishment of pathogenic bacteria within the intestine.

Saccharomyces

Saccharomyces is a genus in the kingdom of fungi that includes many species of yeast. Saccharomyces is from Latin meaning sugar fungi. Many members of this genus are considered very important in food production.

Saccharomyces cerevisiae	Saccharomyces cerevisiae is a species of budding yeast. It is perhaps the most useful yeast owing to its use since ancient times in baking and brewing. It is believed that it was originally isolated from the skins of grapes
Inclusions	Inclusions are chemical substances that may or may not be present in a cell, depending on the cell type. Inclusions are stored nutrients, secretory products, and pigment granules. Examples of Inclusions are; glycogen granules in the liver and muscle cells; lipid droplets in fat cells; pigment granules in certain cells of skin and hair; water containing vacuoles; and crystals of various types.
Column	The Column is a reproductive structure that can be found in several plant families: Aristolochiaceae, Orchidaceae, and Stylidiaceae. It is derived from the fusion of both male and female parts (stamens and pistil) into a single organ. This means that the style and stigma of the pistil, with the filaments and one or more anthers, are all united.
Gene	A Gene is the basic unit of heredity in a living organism. All living things depend on Gene s. Gene s hold the information to build and maintain their cells and pass Gene tic traits to offspring.
Miller	A Miller usually refers to a person who operates a mill, a machine to grind a cereal crop to make flour. Milling is among the oldest of human occupations. "Miller", "Milne" and other variants are common surnames, as are their equivalents in other languages around the world .
Edward	The Edward mango is a monoembryonic mango cultivar grown predominantly in Florida. It is considered by many to be among the finest tasting mangoes in the world; however, its poor yields have restrained the Edward from developing into a commercially significant variety. The Edward was first propagated in the 1920s by Edward Simmonds of the Plant Introduction Garden in Miami, Florida and is believed to be a hybrid cross of Haden and Carabao mango cultivars.

Charles Robert Darwin

Charles Robert Darwin FRS (12 February 1809 - 19 April 1882) was an English naturalist who realised and presented compelling evidence that all species of life have evolved over time from common ancestors, through the process he called natural selection. The fact that evolution occurs became accepted by the scientific community and much of the general public in his lifetime, while his theory of natural selection came to be widely seen as the primary explanation of the process of evolution in the 1930s, and now forms the basis of modern evolutionary theory. In modified form, Darwin"s scientific discovery is the unifying theory of the life sciences, providing logical explanation for the diversity of life.

Species

In biology, a Species is:

· a taxonomic rank (the basic rank of biological classification) or
· a unit at that rank

There are many definitions of what kind of unit a Species is (or should be.) A common definition is that of a group of organisms capable of interbreeding and producing fertile offspring, and separated from other such groups with which interbreeding does not (normally) happen. Other definitions may focus on similarity of DNA or morphology. Some Species are further subdivided into sub Species , and here also there is no close agreement on the criteria to be used.

Animals

Animals are a major group of mostly multicellular, eukaryotic organisms of the kingdom Animalia or Metazoa. Their body plan eventually becomes fixed as they develop, although some undergo a process of metamorphosis later on in their life. Most Animals are motile, meaning they can move spontaneously and independently.

Antibiotic resistance

Antibiotic resistance is the ability of a microorganism to withstand the effects of antibiotics. It is a specific type of drug resistance. antibiotic resistance evolves via natural selection acting upon random mutation, but it can also be engineered by applying an evolutionary stress on a population.

Artificial selection

Artificial selection describes intentional breeding for certain traits, or combination of traits. It was defined by Charles Darwin in contrast to natural selection, in which the differential reproduction of organisms with certain traits is attributed to improved survival or reproductive ability ("Darwinian fitness") and the creation of new organisms, us humans breed them for specific traits. Artificial selection can also be unintentional; it is thought that domestication of crops by early humans was largely unintentional.

Plants

Plants are living organisms belonging to the kingdom Plantae. They include familiar organisms such as trees, herbs, bushes, grasses, vines, ferns, mosses, and green algae. About 350,000 species of Plants, defined as seed Plants, bryophytes, ferns and fern allies, are estimated to exist currently.

Domestication	Domestication or taming refers to the process whereby a population of animals or plants, through a process of selection, becomes accustomed to human provision and control. A defining characteristic of Domestication is artificial selection by humans. Some species such as the Asian Elephant, numerous members of which which have for many centuries been used as working animals, are not domesticated because they have not normally been bred under human control, even though they have been commonly tamed.
Gene	A Gene is the basic unit of heredity in a living organism. All living things depend on Gene s. Gene s hold the information to build and maintain their cells and pass Gene tic traits to offspring.
Broccoli	Broccoli is a plant of the cabbage family Brassicaceae (formerly Cruciferae.) It is classified as a cultivar group of the species Brassica oleracea. Broccoli possesses abundant arboreal, luscious, fleshy, flower heads, usually green in color, arranged in a tree-like fashion on branches sprouting from a thick, edible, sturdy, meaty stalk.
Cauliflower	Cauliflower is one of several vegetables in the species Brassica oleracea, in the family Brassicaceae. It is an annual plant that reproduces by seed. Typically, only the head (the white curd) is eaten while the stalk and surrounding thick, green leaves are used in vegetable broth or discarded.
Cabbage	The Cabbage is a popular cultivar of a the species Brassica oleracea Linne (Capitata Group) of the Family Brassicaceae (or Cruciferae), and is used as a leafy green vegetable. It is a herbaceous, biennial, dicotyledonous flowering plant distinguished by a short stem upon which is crowded a mass of leaves, usually green but in some varieties red or purplish, which while immature form a characteristic compact, globular cluster (Cabbage head.) The plant is also called head Cabbage or heading Cabbage and in Scotland a bowkail, from its rounded shape.
Sprouts	Sprouts is a privately owned Arizona-based chain of specialty grocery stores with locations in Arizona, California, Colorado and Texas. Sprouts stores are smaller than a typical supermarket, focusing on fresh foods, produce, healthy foods and vitamins/supplements. Management hopes to have over 50 stores by 2010.
Adaptation	In ocular physiology, Adaptation is the ability of the eye to adjust to various levels of darkness and light. The human eye can function from very dark to very bright levels of light -- its sensing capabilities reach across nine orders of magnitude. This means that the brightest and the darkest light signal that the eye can sense are a factor of roughly one thousand million apart.
Altruism	Altruism is a well-documented animal behaviour, which appears most obviously in kin relationships but may also be evident amongst wider social groups.

	In the science of ethology (the study of behavior), and more generally in the study of social evolution, Altruism refers to behavior by an individual that increases the fitness of another individual while decreasing the fitness of the actor. Research in evolutionary theory has been applied to social behaviour, including Altruism.
Flower	A Flower sometimes known as a bloom or blossom, is the reproductive structure found in Flower ing plants The biological function of a Flower is to mediate the union of male sperm with female ovum in order to produce seeds. The process begins with pollination, is followed by fertilization, leading to the formation and dispersal of the seeds.
Genetic	Genetics is the study of how living things receive common traits from previous generations. These traits are described by the Genetic information carried by a molecule called DNA. The instructions for constructing and operating an organism are contained in the organism"s DNA. Every living thing on earth has DNA in its cells. Genes are the hereditary components of DNA that occupy spots on chromosomes and determine characteristics in an organism.
Genetic variation	Genetic variation, variation in alleles of genes, occurs both within and among populations. genetic variation is important because it provides the "raw material" for natural selection. genetic variation among individuals within a population can be identified at a variety of levels.
Heritability	In genetics, heritability is the proportion of phenotypic variation in a population that is attributable to genetic variation among individuals. Variation among individuals may be due to genetic and/or environmental factors. heritability analyses estimate the relative contributions of differences in genetic and non-genetic factors to the total phenotypic variance in a population.
Rosemary	Rosemary is a woody, perennial herb with fragrant evergreen needle-like leaves. It is native to the Mediterranean region. It is a member of the mint family Lamiaceae, which also includes many other herbs.
Beak	The Beak bill or rostrum is an external anatomical structure of birds which is used for eating and for grooming, manipulating objects, killing prey, probing for food, courtship and feeding young. The term also refers to a similar mouthpart in some monotremes, cephalopods, cetaceans, pufferfishes, turtles, Anuran tadpoles and sirens. Comparison of bird Beak s, displaying different shapes adapted to different feeding methods.
Population	In biology, a population is the collection of inter-breeding organisms of a particular species; in sociology, a collection of human beings. Individuals within a population share a factor may be reduced by statistical means, but such a generalization may be too vague to imply anything. Demography is used extensively in marketing, which relates to economic units, such as retailers, to potential customers.
Hollywood	A Hollywood is a drug slang term referring to an unusually large and often dangerous single dosage of cocaine.

Depending on a user''s experience with and tolerance of the drug, a typical line measures between 50-75mg of cocaine. However, a Hollywood line can exceed 200 mg of cocaine in a single dosage.

Maternal effect

A Maternal effect is a situation where the phenotype of an organism is determined not only by the environment it experiences and its genotype, but also by the environment and phenotype of its mother. In genetics, Maternal effect s occur when an organism shows the phenotype expected from the genotype of the mother, irrespective of its own genotype, often due the mother supplying mRNA or proteins to the egg. Maternal effect s can also be caused by the maternal environment independent of genotype, sometimes controlling the size, sex, or behaviour of the offspring.

Nest

A Nest is a place of refuge to hold an animal''s eggs and/or provide a place to live or raise offspring. They are usually made of some organic material such as twigs, grass, and leaves; or may simply be a depression in the ground, or a hole in a tree, rock or building. Human-made materials, such as string, plastic, cloth, hair or paper, may be used.

Protein

Protein s are organic compounds made of amino acids arranged in a linear chain. The amino acids in a polymer chain are joined together by the peptide bonds between the carboxyl and amino groups of adjacent amino acid residues. The sequence of amino acids in a protein is defined by the sequence of a gene, which is encoded in the genetic code.

Allele

An Allele) is one of a series of different forms of a gene. The word is a short from of allelomorph , which was used in the early days of genetics to describe variant forms of a gene detected as different phenotypes. Allele s are now understood to be alternative DNA sequences at the same physical gene locus, which may or may not result in different phenotypic traits.

Allele frequency

Allele frequency is the proportion of all variants (alleles) that is made up of a particular gene variant. In other words, it is the number of copies of a particular allele divided by the number of copies of all alleles at the genetic place (locus) in a population. It can be expressed for example as a percentage.

Phenotype

A Phenotype is any observable characteristic or trait of an organism: such as its morphology, development, biochemical or physiological properties, or behavior. Phenotype s result from the expression of an organism''s genes as well as the influence of environmental factors and possible interactions between the two. The genotype of an organism is the inherited instructions it carries within its genetic code.

Blending inheritance

In Darwin''s time, biologists held to the theory of Blending inheritance--an offspring was an average of its parent. If an individual had one short parent and one tall parent, it would always be of some interim height between that of its parents. In turn, this interim height would become a theoretical extreme (either tall or short) that, in turn, bounded the potential height of its own offspring.

Trait	A trait is a distinct variant of a phenotypic character of an organism that may be inherited, environmentally determined or somewhere in between. For example, eye color is a character or abstraction of an attribute, while blue, brown and hazel are traits. A trait may be any single feature or quantifiable measurement of an organism.
Anatomy	The anatomy of spiders is in some aspects similar, but also different from that of other arthropods. The following characteristics are common to all spiders: A body with two segments, eight legs, spinnerets, no chewing parts, no wings, and the presence of chelicerae, which spiders use to hold prey, and in most cases, inject venom. Spiders have non-compound eyes, with most species having eight; the spiders known as Haplogynae may have six or fewer, and certain cave-dwelling spiders may have none at all.
Fish	FISH (fluorescence in situ hybridization) is a cytogenetic technique used to detect and localize the presence or absence of specific DNA sequences on chromosomes. FISH uses fluorescent probes that bind to only those parts of the chromosome with which they show a high degree of sequence similarity. Fluorescence microscopy can be used to find out where the fluorescent probe bound to the chromosomes.
Mutation	In biology, mutation s are changes to the nucleotide sequence of the genetic material of an organism. mutation s can be caused by copying errors in the genetic material during cell division, by exposure to ultraviolet or ionizing radiation, chemical mutagens, or viruses, or can be induced by the organism itself, by cellular processes such as hyper mutation . In multicellular organisms with dedicated reproductive cells, mutation s can be subdivided into germ line mutation s, which can be passed on to descendants through the reproductive cells, and somatic mutation s, which involve cells outside the dedicated reproductive group and which are not usually transmitted to descendants.
Leukemia inhibitory factor	Leukemia inhibitory factor an interleukin 6 class cytokine, is a chemical in cells that affects their growth and development. Leukemia inhibitory factor derives its name from its ability to induce the terminal differentiation of myeloid leukaemic cells. Other properties attributed to the cytokine include: the growth promotion and cell differentiation of different types of target cells, influence on bone metabolism, cachexia, neural development, embryogenesis and inflammation.
Edward	The Edward mango is a monoembryonic mango cultivar grown predominantly in Florida. It is considered by many to be among the finest tasting mangoes in the world; however, its poor yields have restrained the Edward from developing into a commercially significant variety. The Edward was first propagated in the 1920s by Edward Simmonds of the Plant Introduction Garden in Miami, Florida and is believed to be a hybrid cross of Haden and Carabao mango cultivars.
Pigment	A pigment is the material that changes the color of light it reflects as the result of selective color absorption. This physical process differs from fluorescence, phosphorescence, and other forms of luminescence, in which the material itself emits light.

Many materials selectively absorb certain wavelengths of light.

Axoneme

Numerous eukaryotic cells carry whip-like appendages (cilia or eukaryotic flagella) whose inner core consists of a cytoskeletal structure called the Axoneme. The Axoneme serves as the "skeleton" of these organelles, both giving support to the structure and, in some cases, causing it to bend. Though distinctions of function and/or length may be made between cilia and flagella, the internal structure of the Axoneme is common to both.

Chlamydomonas

Chlamydomonas is a genus of green alga. They are unicellular flagellates. Chlamydomonas is used as a model organism for molecular biology, especially studies of flagellar motility and chloroplast dynamics, biogenesis, and genetics.

Flagellum

A Flagellum is a tail-like structure that projects from the cell body of certain prokaryotic and eukaryotic cells, and functions in locomotion. There are some notable differences between prokaryotic and eukaryotic flagella, such as protein composition, structure, and mechanism of propulsion. An example of a flagellated bacterium is the ulcer-causing Helicobacter pylori, which uses multiple flagella to propel itself through the mucus lining to reach the stomach epithelium.

Cilium

A Cilium is an organelle found in eukaryotic cells. Cilia are tail-like projections extending approximately 5-10 micrometers from the cell body.

There are two types of cilia: motile cilia and non-motile, or primary, cilia, which typically serve as sensory organelles.

Electrophoresis

Electrophoresis is the best-known electrokinetic phenomenon. It was discovered by Reuss in 1807. He observed that clay particles dispersed in water migrate under influence of an applied electric field.

Antifreeze proteins

Antifreeze proteins or ice structuring proteins (ISPs) refer to a class of polypeptides produced by certain vertebrates, plants, fungi and bacteria that permit their survival in subzero environments. AFPs bind to small ice crystals to inhibit growth and recrystallization of ice that would otherwise be fatal. There is also increasing evidence that AFPs interact with mammalian cell membranes to protect from cold damage.

Crystallin

In biology, a Crystallin is a water-soluble structural protein found in the lens of the eye, accounting for the transparency of the structure and also in cornea, when it can play a similar role. It has also been identified in other places such as the heart and aggressive breast cancer tumors . Since it has been shown that lens injury may promote nerve regeneration, Crystallin has been an area of neural research.

Chromosome

A Chromosome is an organized structure of DNA and protein that is found in cells. It is a single piece of coiled DNA containing many genes, regulatory elements and other nucleotide sequences. Chromosome s also contain DNA-bound proteins, which serve to package the DNA and control its functions.

Inversion	An inversion is a chromosome rearrangement in which a segment of a chromosome is reversed end to end. An inversion occurs when a single chromosome undergoes breakage and rearrangement within itself. Inversions are of two types: paracentric and pericentric.
Fentanyl	Fentanyl -- brand names include Actiq, Duragesic, Fentora, and Sublimaze -- is a synthetic primary μ-opioid agonist commonly used to treat post-operative and chronic breakthrough pain. It is approximately 100 times more potent than morphine, with 100 micrograms of Fentanyl equivalent to 10 mg of morphine and 75 mg of pethidine in analgesic activity. It has an LD_{50} of 3.1 milligrams per kilogram in rats, 0.03 milligrams per kilogram in monkeys, and an undetermined milligrams per kilogram in humans.
David	DAVID (the Database for Annotation, Visualization and Integrated Discovery) is a free online bioinformatics resources developed by the Laboratory of Immunopathogenesis and Bioinformatics (LIB.) All tools in the DAVID Bioinformatics Resources aim to provide functional interpretation of large lists of genes derived from genomic studies, e.g. microarray and proteomics studies. DAVID can be found at http://DAVID.niaid.nih.gov or http://DAVID.abcc.ncifcrf.gov The newly updated DAVID Bioinformatics Resources consists of the DAVID Knowledgebase and five integrated, web-based functional annotation tool suites: the DAVID Gene Functional Classification Tool, the DAVID Functional Annotation Tool, the DAVID Gene ID Conversion Tool, the DAVID Gene Name Viewer and the DAVID NIAID Pathogen Genome Browser.

Trait	A trait is a distinct variant of a phenotypic character of an organism that may be inherited, environmentally determined or somewhere in between. For example, eye color is a character or abstraction of an attribute, while blue, brown and hazel are traits. A trait may be any single feature or quantifiable measurement of an organism.
Group	In naming cultivated plants, a Group is a formal classification category, under the International Code of Nomenclature for Cultivated Plants (ICNCP): ICNCP Art. 3.1: "a formal category for assembing cultivars, individual plants or assemblages of plants on the basis of defined similarity" . The term "Group" (with a capital G) was introduced in the 2004 ICNCP, replacing the "Cultivar-Group" of the 1995 ICNCP. A Group is united by some common trait; for example there may be a Group of yellow-flowering cultivars, a Group of cultivars with variegated leaves, a Group of cultivars resistant to a particular disease, etc.
Heritability	In genetics, heritability is the proportion of phenotypic variation in a population that is attributable to genetic variation among individuals. Variation among individuals may be due to genetic and/or environmental factors. heritability analyses estimate the relative contributions of differences in genetic and non-genetic factors to the total phenotypic variance in a population.
Camerae	Camerae are the spaces or chambers enclosed between two adjacent septa in the phragmocone of a nautiloid or ammonoid cephalopod. These can be seen in cross-sections of a nautilus shell and in the polished cross-sections of ammonites. In life these chambers are filled with gas, mediated by the siphuncle, and used to control buoyancy.
Octopus	The Octopus is a cephalopod of the order Octopoda. The Octopus inhabits many diverse regions of the ocean, especially coral reefs. The term may also be used to refer only to those creatures in the genus Octopus.
Fish	FISH (fluorescence in situ hybridization) is a cytogenetic technique used to detect and localize the presence or absence of specific DNA sequences on chromosomes. FISH uses fluorescent probes that bind to only those parts of the chromosome with which they show a high degree of sequence similarity. Fluorescence microscopy can be used to find out where the fluorescent probe bound to the chromosomes.
DNA	Deoxyribonucleic acid (DNA) is a nucleic acid that contains the genetic instructions used in the development and functioning of all known living organisms and some viruses. The main role of DNA molecules is the long-term storage of information. DNA is often compared to a set of blueprints or a recipe, or a code, since it contains the instructions needed to construct other components of cells, such as proteins and RNA molecules.
DNA sequence	A DNA sequence or genetic sequence is a succession of letters representing the primary structure of a real or hypothetical DNA molecule or strand, with the capacity to carry information as described by the central dogma of molecular biology.

	The possible letters are A, C, G, and T, representing the four nucleotide bases of a DNA strand -- adenine, cytosine, guanine, thymine -- covalently linked to a phosphodiester backbone. In the typical case, the sequences are printed abutting one another without gaps, as in the sequence AAAGTCTGAC, read left to right in the 5" to 3" direction.
Animals	Animals are a major group of mostly multicellular, eukaryotic organisms of the kingdom Animalia or Metazoa. Their body plan eventually becomes fixed as they develop, although some undergo a process of metamorphosis later on in their life. Most Animals are motile, meaning they can move spontaneously and independently.
Gene	A Gene is the basic unit of heredity in a living organism. All living things depend on Gene s. Gene s hold the information to build and maintain their cells and pass Gene tic traits to offspring.
Genetic	Genetics is the study of how living things receive common traits from previous generations. These traits are described by the Genetic information carried by a molecule called DNA. The instructions for constructing and operating an organism are contained in the organism"s DNA. Every living thing on earth has DNA in its cells. Genes are the hereditary components of DNA that occupy spots on chromosomes and determine characteristics in an organism.
David	DAVID (the Database for Annotation, Visualization and Integrated Discovery) is a free online bioinformatics resources developed by the Laboratory of Immunopathogenesis and Bioinformatics (LIB.) All tools in the DAVID Bioinformatics Resources aim to provide functional interpretation of large lists of genes derived from genomic studies, e.g. microarray and proteomics studies. DAVID can be found at http://DAVID.niaid.nih.gov or http://DAVID.abcc.ncifcrf.gov The newly updated DAVID Bioinformatics Resources consists of the DAVID Knowledgebase and five integrated, web-based functional annotation tool suites: the DAVID Gene Functional Classification Tool, the DAVID Functional Annotation Tool, the DAVID Gene ID Conversion Tool, the DAVID Gene Name Viewer and the DAVID NIAID Pathogen Genome Browser.
Hollywood	A Hollywood is a drug slang term referring to an unusually large and often dangerous single dosage of cocaine. Depending on a user"s experience with and tolerance of the drug, a typical line measures between 50-75mg of cocaine. However, a Hollywood line can exceed 200 mg of cocaine in a single dosage.

Order

In scientific classification used in biology, the Order is

- a taxonomic rank used in the classification of organisms. Other well-known ranks are life, domain, kingdom, phylum, class, family, genus, and species, with Order fitting in between class and family. An immediately higher rank, superorder, may be added directly above Order, while suborder would be a lower rank.
- a taxonomic unit, a taxon, in that rank. In that case the plural is orders .

The Latin suffix -formes meaning "having the form of" is used for the scientific name of orders of birds and reptiles, but not for those of mammals and invertebrates.

The Order as a distinct rank of biological classification having its own distinctive name (and not just called a higher genus (genus summum)) was first introduced by a German botanist Augustus Quirinus Rivinus in his classification of plants . Carolus Linnaeus was the first to apply it consistently to the division of all three kingdoms of Nature (minerals, plants, and animals) in his Systema Naturae (1735, 1st. Ed.).

Plants

Plants are living organisms belonging to the kingdom Plantae. They include familiar organisms such as trees, herbs, bushes, grasses, vines, ferns, mosses, and green algae. About 350,000 species of Plants, defined as seed Plants, bryophytes, ferns and fern allies, are estimated to exist currently.

Clock

CLOCK, or CLOCK is a gene which encodes proteins regulating circadian rhythm. The CLOCK protein seems to affect both the persistence and length of the circadian cycle. CLOCK forms part of a basic-helix-loop-helix transcription factor.

Genetic drift

Genetic drift or allelic drift is the change in the relative frequency with which a gene variant (allele) occurs in a population that results from the fact that alleles in offspring are a random sample of those in the parents, and because of the role of chance in determining whether a given individual survives and reproduces. A population"s allele frequency is the fraction of the gene copies that share a particular form.

Genetic drift is one of several evolutionary processes which lead to changes in allele frequencies over time.

Heads

The capitulum is considered the most derived form of inflorescence. Flower Heads found outside Asteraceae (such as those of Eryngium or Jasione) shows lesser degrees of specialization. .

Lice

Lice, are an order of over 3,000 species of wingless insects; three of which are classified as human disease agents. They are obligate ectoparasites of every avian and most mammalian orders. They are not found on Monotremes and a few eutherian orders, namely the bats, whales, dolphins and porpoises and pangolins

Molecular clock	The Molecular clock) is a technique in molecular evolution to relate the time that two species diverged to the number of molecular differences measured between the species" DNA sequences or proteins. It is sometimes called a gene clock or evolutionary clock. The notion of the existence of a so-called "Molecular clock" was first attributed to Emile Zuckerkandl and Linus Pauling who, in 1962, noticed that the number of amino acid differences in hemoglobin between different lineages roughly changes with time, as estimated also from fossil evidence.
Neutral theory of molecular evolution	The Neutral theory of molecular evolution is an influential theory, which was introduced with effect by Motoo Kimura in the late 1960s and early 1970s. The theory states that the vast majority of evolutionary changes at the molecular level are caused by random drift of selectively neutral mutants. Although the theory was received by some as an argument against Darwin"s theory of evolution by natural selection, Kimura maintained, and most modern evolutionary biologists agree, that the two theories are compatible: "The theory does not deny the role of natural selection in determining the course of adaptive evolution" (Kimura, 1986.)
Rat	Rat s are various medium-sized, long-tailed rodents of the superfamily Muroidea. "True Rat s" are members of the genus Rat tus, the most important of which to humans are the black Rat Rat tus Rat tus, and the brown Rat Rat tus norvegicus. Many members of other rodent genera and families are also called Rat s and share many characteristics with true Rat s.
Biodiversity	Biodiversity is the variation of life forms within a given ecosystem, biome, or for the entire Earth. biodiversity is often used as a measure of the health of biological systems. The biodiversity found on Earth today consists of many millions of distinct biological species, which is the product of nearly 3.5 billion years of evolution.
Chameleon	The family Chamaeleonidae are a distinctive and highly specialized clade of lizards. They are distinguished by their parrot-like zygodactylous feet, their separately mobile and stereoscopic eyes, their very long, highly modified, and rapidly extrudable tongues, their swaying gait, and the possession by many of a prehensile tail, crests or horns on their distinctively shaped heads, and the ability of some to change color. Uniquely adapted for climbing and visual hunting, the approximately 160 species of Chameleon range from Africa, Madagascar, Spain and Portugal, across south Asia, to Sri Lanka, have been introduced to Hawaii and California, and are found in warm habitats that vary from rain forest to desert conditions.
Adaptation	In ocular physiology, Adaptation is the ability of the eye to adjust to various levels of darkness and light. The human eye can function from very dark to very bright levels of light -- its sensing capabilities reach across nine orders of magnitude. This means that the brightest and the darkest light signal that the eye can sense are a factor of roughly one thousand million apart.

Bacteriocyte	A Bacteriocyte is a specialized adipocyte found in some insect groups such as aphids, tsetse flies, german cockroaches, and many others. Mycetocyte symbionts specifically include three insect groups: Cockroaches, Homoptera such as aphid, and Coleoptera such as weevils. These cells contain endosymbiotic bacteria such as Buchnera species, which provide essential amino acids and other chemicals to their host.
Coevolution	In a broad sense, biological coevolution is "the change of a biological object triggered by the change of a related object". coevolution can occur at multiple levels of biology: it can be as microscopic as correlated mutations between amino acids in a protein, or as macroscopic as covarying traits between different species in an environment. Each party in a coevolution ary relationship exerts selective pressures on the other, thereby affecting each others" evolution.
Buchnera	Buchnera aphidicola a member of the Proteobacteria, is the primary endosymbiont of aphids (A. pisum.) It is believed that Buchnera was once a free living gram negative ancestor similar to a modern Enterobacteriaceae such as Escherichia coli. Buchnera are 3 Âµm in diameter and have some of the key characteristics of their Enterobacteriaceae relatives such as a gram-negative cell wall.
Mammal	Mammal s (formally mammal ia) are a class of vertebrate animals whose females are characterized by the possession of mammary glands while both males and females are characterized by sweat glands, hair, three middle ear bones used in hearing, and a neocortex region in the brain. mammal s are divided into three main categories depending how they are born. These categories are, monotremes, marsupials and placentals.
Ant	Ants are social insects of the family Formicidae , and along with the related wasps and bees, they belong to the order Hymenoptera. Ants evolved from wasp-like ancestors in the mid-Cretaceous period between 110 and 130 million years ago and diversified after the rise of flowering plants. Today, more than 12,000 species are classified with upper estimates of about 14,000 species.
Fungi	A fungus is a eukaryotic organism that is a member of the kingdom fungi . The fungi are a monophyletic group, also called the Eumycota , that is phylogenetically distinct from the structurally similar slime molds (myxomycetes) and water molds (oomycetes.) The fungi are heterotrophic organisms possessing a chitinous cell wall, with most species growing as multicellular filaments called hyphae forming a mycelium; some species also grow as single cells.
Drosophila	Drosophila has long been a favorite model system for geneticists and developmental biologists studying embryogenesis. The small size, short generation time, and large brood size makes it ideal for genetic studies. Transparent embryos facilitate developmental studies.

Hox genes	Hox genes are a group of related genes that specify the anterior-posterior axis and segment identity of metazoan organisms during early embryonic development. These genes are critical for the proper number and placement of embryonic segment structures (such as legs, antennae, and eyes.) Homeotic genes are defined by a DNA sequence known as the homeobox, which is a sequence of 180 nucleotides that code for a protein domain known as the homeodomain.
Mutant	A Mutant is an individual, organism which is a base-pair sequence change within the DNA of a gene or chromosome of an organism resulting in the creation of a new character or trait not found in the wild type. The natural occurrence of genetic mutations is integral to the process of evolution. The study of Mutant s is an integral part of biology, by understanding the effect that a mutation in a gene has it is possible to establish the normal function of that gene.

Mutation	In biology, mutation s are changes to the nucleotide sequence of the genetic material of an organism. mutation s can be caused by copying errors in the genetic material during cell division, by exposure to ultraviolet or ionizing radiation, chemical mutagens, or viruses, or can be induced by the organism itself, by cellular processes such as hyper mutation . In multicellular organisms with dedicated reproductive cells, mutation s can be subdivided into germ line mutation s, which can be passed on to descendants through the reproductive cells, and somatic mutation s, which involve cells outside the dedicated reproductive group and which are not usually transmitted to descendants.
Allele	An Allele) is one of a series of different forms of a gene. The word is a short from of allelomorph , which was used in the early days of genetics to describe variant forms of a gene detected as different phenotypes. Allele s are now understood to be alternative DNA sequences at the same physical gene locus, which may or may not result in different phenotypic traits.
DNA	Deoxyribonucleic acid (DNA) is a nucleic acid that contains the genetic instructions used in the development and functioning of all known living organisms and some viruses. The main role of DNA molecules is the long-term storage of information. DNA is often compared to a set of blueprints or a recipe, or a code, since it contains the instructions needed to construct other components of cells, such as proteins and RNA molecules.
DNA structure	DNA structure shows a variety of forms, both double-stranded and single-stranded. The mechanical properties of DNA, which are directly related to its structure, are a significant problem for cells. Every process which binds or reads DNA is able to use or modify the mechanical properties of DNA for purposes of recognition, packaging and modification.
Deoxyribonucleotide	A Deoxyribonucleotide is the monomer of DNA a deoxyribose sugar, and one or more phosphate groups. The nitrogenous base is always bonded to the 1" carbon of the deoxyribose, which is distinguished from ribose by the presence of a proton on the 2" carbon rather than an -OH group.
Deoxyribose	Deoxyribose, also known as D-Deoxyribose and 2-Deoxyribose, is an aldopentose -- a monosaccharide containing five carbon atoms, and including an aldehyde functional group in its linear structure. It is a deoxy sugar derived from the pentose sugar ribose by the replacement of the hydroxyl group at the 2 position with hydrogen, leading to the net loss of an oxygen atom. Replacement of the hydroxyl group also shifts the conformation of the ring from C3"-endo to C2"-endo.
Phosphate	A Phosphate an inorganic chemical, is a salt of phosphoric acid. Inorganic Phosphate s are mined to obtain phosphorus for use in agriculture and industry. In organic chemistry, a Phosphate or organo Phosphate , is an ester of phosphoric acid.

Phosphodiester bond	A Phosphodiester bond is a group of strong covalent bonds between the phosphorus atom in a phosphate group and two other molecules over two ester bonds. Phosphodiester bond s are central to all life on Earth, as they make up the backbone of the strands of DNA. In DNA and RNA, the Phosphodiester bond is the linkage between the 3" carbon atom of one sugar molecule and the 5" carbon of another, deoxyribose in DNA and ribose in RNA. The phosphate groups in the Phosphodiester bond are negatively-charged. Because the phosphate groups have a pK_a near 0, they are negatively-charged at pH 7.
Pyrimidine	Pyrimidine is a heterocyclic aromatic organic compound similar to benzene and pyridine, containing two nitrogen atoms at positions 1 and 3 of the six-member ring. It is isomeric with two other forms of diazine. Three nucleobases found in nucleic acids (cytosine, thymine, and uracil) are Pyrimidine derivatives: In DNA and RNA, these bases form hydrogen bonds with their complementary purines.
Base pair	In molecular biology, two nucleotides on opposite complementary DNA or RNA strands that are connected via hydrogen bonds are called a Base pair In the canonical Watson-Crick Base pair ing, adenine (A) forms a Base pair with thymine (T), as does guanine (G) with cytosine (C) in DNA. In RNA, thymine is replaced by uracil (U.) Non-Watson-Crick Base pair ing with alternate hydrogen bonding patterns also occur, especially in RNA; common such patterns are Hoogsteen Base pair s.
Group	In naming cultivated plants, a Group is a formal classification category, under the International Code of Nomenclature for Cultivated Plants (ICNCP): ICNCP Art. 3.1: "a formal category for assembing cultivars, individual plants or assemblages of plants on the basis of defined similarity" . The term "Group" (with a capital G) was introduced in the 2004 ICNCP, replacing the "Cultivar-Group" of the 1995 ICNCP. A Group is united by some common trait; for example there may be a Group of yellow-flowering cultivars, a Group of cultivars with variegated leaves, a Group of cultivars resistant to a particular disease, etc.
DNA polymerase	A DNA polymerase is an enzyme that catalyzes the polymerization of deoxyribonucleotides into a DNA strand. DNA polymerase s are best-known for their role in DNA replication, in which the polymerase "reads" an intact DNA strand as a template and uses it to synthesize the new strand. The newly-polymerized molecule is complementary to the template strand and identical to the template"s original partner strand.
Gene	A Gene is the basic unit of heredity in a living organism. All living things depend on Gene s. Gene s hold the information to build and maintain their cells and pass Gene tic traits to offspring.

Hemoglobin	Hemoglobin is the iron-containing oxygen-transport metalloprotein in the red blood cells of vertebrates, and the tissues of some invertebrates. In mammals, the protein makes up about 97% of the red blood cell"s dry content, and around 35% of the total content . Hemoglobin transports oxygen from the lungs or gills to the rest of the body where it releases the oxygen for cell use.
Messenger ribonucleic acid	Messenger ribonucleic acid is a molecule of RNA encoding a chemical "blueprint" for a protein product. mRNA is transcribed from a DNA template, and carries coding information to the sites of protein synthesis: the ribosomes. Here, the nucleic acid polymer is translated into a polymer of amino acids: a protein.
Anemia	Species of the genus Anemia are sometimes called flowering ferns, but this term is more commonly applied to ferns of the genus Osmunda. It is sometimes classified in family Schizaeaceae. · A. adiantifolia - pineland fern · A. cicutaria - hemlock fern · A. coriacia · A. cuneata · A. donnell-smithii · A. hirsuta - hairy flowering fern · A. hirta - streambank flowering fern · A. mexicana - Mexican flowering fern · A. phyllitidis · A. portoricensis - Puerto Rican flowering fern · A. underwoodiana · A. wrightii - Wright"s flowering fern .
Chromosome	A Chromosome is an organized structure of DNA and protein that is found in cells. It is a single piece of coiled DNA containing many genes, regulatory elements and other nucleotide sequences. Chromosome s also contain DNA-bound proteins, which serve to package the DNA and control its functions.
Inversion	An inversion is a chromosome rearrangement in which a segment of a chromosome is reversed end to end. An inversion occurs when a single chromosome undergoes breakage and rearrangement within itself. Inversions are of two types: paracentric and pericentric.
Enzymes	Enzymes are biomolecules that catalyze (i.e., increase the rates of) chemical reactions. Nearly all known Enzymes are proteins. However, certain RNA molecules can be effective biocatalysts too.
Achondroplasia	Achondroplasia dwarfism is a type of autosomal dominant genetic disorder that is a common cause of dwarfism. Achondroplastic dwarfs have short stature, with an average adult height of 131 cm (4 feet, 3-1/2 inches) for males and 123 cm (4 feet, 1/2 inches) for females.

The prevalence is approximately 1 in 25,000.

Genetic

Genetics is the study of how living things receive common traits from previous generations. These traits are described by the Genetic information carried by a molecule called DNA. The instructions for constructing and operating an organism are contained in the organism"s DNA. Every living thing on earth has DNA in its cells. Genes are the hereditary components of DNA that occupy spots on chromosomes and determine characteristics in an organism.

Genetic variation

Genetic variation, variation in alleles of genes, occurs both within and among populations. genetic variation is important because it provides the "raw material" for natural selection.

genetic variation among individuals within a population can be identified at a variety of levels.

Nematode

The "roundworms" or " Nematode s" (phylum Nematoda) are the most diverse phylum of pseudocoelomates, and one of the most diverse of all animals. Nematode species are very difficult to distinguish; over 80,000 have been described, of which over 15,000 are parasitic. It has been estimated that the total number of described and undescribed roundworms might be more than 500,000.

Heritability

In genetics, heritability is the proportion of phenotypic variation in a population that is attributable to genetic variation among individuals. Variation among individuals may be due to genetic and/or environmental factors. heritability analyses estimate the relative contributions of differences in genetic and non-genetic factors to the total phenotypic variance in a population.

Rat

Rat s are various medium-sized, long-tailed rodents of the superfamily Muroidea. "True Rat s" are members of the genus Rat tus, the most important of which to humans are the black Rat Rat tus Rat tus, and the brown Rat Rat tus norvegicus. Many members of other rodent genera and families are also called Rat s and share many characteristics with true Rat s.

Worm

The term Worm is used to describe many different distantly-related animals which have a long cylindrical body and no legs.

Most animals called " Worm s" are invertebrates, but the term is also used for the amphibian caecilians and the slow Worm Anguis, a legless burrowing lizard. Invertebrate animals commonly called " Worm s" include annelids (earth Worm s), nematodes (round Worm s), flat Worm s, marine polychaete Worm s (bristle Worm s), marine nemertean Worm ("bootlace Worm s"), and insect larvae such as caterpillars, grubs, and maggots.

Adaptation

In ocular physiology, Adaptation is the ability of the eye to adjust to various levels of darkness and light.

The human eye can function from very dark to very bright levels of light -- its sensing capabilities reach across nine orders of magnitude. This means that the brightest and the darkest light signal that the eye can sense are a factor of roughly one thousand million apart.

DNA repair

DNA repair refers to a collection of processes by which a cell identifies and corrects damage to the DNA molecules that encode its genome. In human cells, both normal metabolic activities and environmental factors such as UV light and Radiation can cause DNA damage, resulting in as many as 1 million individual molecular lesions per cell per day. Many of these lesions cause structural damage to the DNA molecule and can alter or eliminate the cell"s ability to transcribe the gene that the affected DNA encodes.

Escherichia

Escherichia is a genus of Gram-negative, non-spore forming, facultatively anaerobic, rod-shaped bacteria from the family Enterobacteriaceae. Inhabitants of the gastrointestinal tracts of warm-blooded animals, Escherichia species provide a portion of the microbially-derived vitamin K for their host.

While many Escherichia are harmless commensals, particular strains of some species are human pathogens, and are known as the most common cause of urinary tract infections, significant sources of gastrointestinal disease, ranging from simple diarrhea to dysentery-like conditions, as well as a wide-range of other pathogenic states.

Escherichia coli

Escherichia coli , is a Gram negative bacterium that is commonly found in the lower intestine of warm-blooded organisms . Most E. coli strains are harmless, but some, such as serotype O157:H7, can cause serious food poisoning in humans, and are occasionally responsible for costly product recalls. The harmless strains are part of the normal flora of the gut, and can benefit their hosts by producing vitamin K_2, or by preventing the establishment of pathogenic bacteria within the intestine.

Indel

Indel is a molecular biology term that has different definitions in different fields:

· In evolutionary studies, Indel is used to mean an insertion or a deletion and Indel s simply refers to the mutation class that includes both insertions, deletions, and the combination thereof, including insertion and deletion events that may be separated by many years.
· In germline and somatic mutation studies, however, Indel describes a special mutation class, defined as a mutation resulting in a colocalized insertion and deletion and a net gain or loss in nucleotides, and micro Indel is defined as an Indel that results in a net gain or loss of 1 to 50 nucleotides.

In coding regions of the genome, unless the length of an Indel is a multiple of 3, they produce a frameshift mutation. For example, a common micro Indel which results in a frameshift causes Bloom syndrome in the Japanese population. Indel s can be contrasted with a point mutation; where an Indel inserts and deletes nucleotides from a sequence, a point mutation is a form of substitution that replaces one of the nucleotides. Indel s can also be contrasted with Tandem Base Mutations , which may result from fundamentally different mechanisms.

Species

In biology, a Species is:

· a taxonomic rank (the basic rank of biological classification) or
· a unit at that rank

	There are many definitions of what kind of unit a Species is (or should be.) A common definition is that of a group of organisms capable of interbreeding and producing fertile offspring, and separated from other such groups with which interbreeding does not (normally) happen. Other definitions may focus on similarity of DNA or morphology. Some Species are further subdivided into sub Species , and here also there is no close agreement on the criteria to be used.
Salmonella	Salmonella is a genus of rod-shaped, Gram-negative, non-spore forming, predominantly motile enterobacteria with diameters around 0.7 to 1.5 Âµm, lengths from 2 to 5 Âµm, and flagella which project in all directions (i.e. peritrichous.) They are chemoorganotrophs, obtaining their energy from oxidation and reduction reactions using organic sources and are facultative anaerobes; most species produce hydrogen sulfide, which can readily be detected by growing them on media containing ferrous sulfate, such as TSI. Most isolates exist in two phases; phase I is the motile phase and phase II the non-motile phase. Cultures that are non-motile upon primary culture may be swithched to the motile phase using a Craigie tube.
Saccharomyces	Saccharomyces is a genus in the kingdom of fungi that includes many species of yeast. Saccharomyces is from Latin meaning sugar fungi. Many members of this genus are considered very important in food production.
Saccharomyces cerevisiae	Saccharomyces cerevisiae is a species of budding yeast. It is perhaps the most useful yeast owing to its use since ancient times in baking and brewing. It is believed that it was originally isolated from the skins of grapes
Gene duplication	Gene duplication is any duplication of a region of DNA that contains a gene; it may occur as an error in homologous recombination, a retrotransposition event, or duplication of an entire chromosome. The second copy of the gene is often free from selective pressure -- that is, mutations of it have no deleterious effects to its host organism. Thus it mutates faster than a functional single-copy gene, over generations of organisms.
Meiosis	In biology, Meiosis is a process of reductional division in which the number of chromosomes per cell is halved. In animals, Meiosis always results in the formation of gametes, while in other organisms it can give rise to spores. As with mitosis, before Meiosis begins, the DNA in the original cell is replicated during S-phase of the cell cycle.
CD36	CD36 is an integral membrane protein found on the surface of many cell types in vertebrate animals and is also known as FAT, SCARB3, GP88, glycoprotein IV and glycoprotein IIIb CD36 is a member of the class B scavenger receptor family of cell surface proteins. CD36 binds many ligands including collagen, thrombospondin, erythrocytes parasitized with Plasmodium falciparum, oxidized low density lipoprotein, native lipoproteins, oxidized phospholipids, and long-chain fatty acids.
Family	In biological classification, Family is

· a taxonomic rank. Other well-known ranks are life, domain, kingdom, phylum, class, order, genus, and species, with Family fitting between order and genus. As for the other well-known ranks, there is the option of an immediately lower rank, indicated by the prefix sub-: subfamily .
· a taxonomic unit, a taxon, in that rank. In that case the plural is families
Example: "Walnuts and Hickories belong to the Walnut Family.
What does and does not belong to each Family is determined by a taxonomist. Similarly for the question if a particular Family should be recognized at all. Often there is no exact agreement, with different taxonomists each taking a different position.

Gene family
A Gene family is a set of genes with a known homology. They are generally biochemically similar. Genes are categorized this way into families, depending on shared nucleotide or protein sequences.

Homeotic genes
Homeotic genes are genes that determine which parts of the body form what body parts. One example is the Hox genes which are important for segmentation, another example are the MADS-box-containing genes in the ABC model of flower development.

Size
Size has been one of the most interesting aspects of cephalopod science to the general public Extinct taxa are also included.

Genetic linkage
Genetic linkage occurs when particular genetic loci or alleles for genes are inherited jointly. Genetic loci on the same chromosome are physically connected and tend to stay together during meiosis, and are thus genetically linked. This is called autosomal linkage.

Cline
In biology, a Cline is a term used to describe a continuum of ecotypes or forms that exhibit gradual phenotypic and/or genetic differences over a geographical area, typically as a result of environmental heterogeneity. Genetically, clines result from the change of allele frequencies within the gene pool of the group of taxa in question. Map of indigenous skin color distribution in the world based on Von Luschan"s chromatic scale.
In the context of anthropology and particularly human genetic variation, clines are related to races and can be genetically defined in a rigorous fashion, for they are readily submissible to straightforward, quantitative metrics.

Chromosomal crossover
Chromosomal crossover is the exchange of genetic material between homologous chromosomes during prophase 1 of meiosis. This often occurs during prophase 1 of meiosis in a process called synapsis. Synapsis begins before the synaptonemal complex develops, and is not completed until near the end of prophase 1.

Drosophila
Drosophila has long been a favorite model system for geneticists and developmental biologists studying embryogenesis. The small size, short generation time, and large brood size makes it ideal for genetic studies. Transparent embryos facilitate developmental studies.

Gene flow	In population genetics, gene flow is the transfer of alleles of genes from one population to another. Migration into or out of a population may be responsible for a marked change in allele frequencies Immigration may also result in the addition of new genetic variants to the established gene pool of a particular species or population.
Plants	Plants are living organisms belonging to the kingdom Plantae. They include familiar organisms such as trees, herbs, bushes, grasses, vines, ferns, mosses, and green algae. About 350,000 species of Plants, defined as seed Plants, bryophytes, ferns and fern allies, are estimated to exist currently.
Polyploidy	Polyploidy occurs in cells and organisms when there are more than two paired (homologous) sets of chromosomes. Known paleo Polyploidy in eukaryotes Most organisms are normally diploid, meaning they have two sets of chromosomes -- one set inherited from each parent. Polyploidy may occur due to abnormal cell division.
Gamete	A Gamete is a cell that fuses with another Gamete during fertilization (conception) in organisms that reproduce sexually. In species that produce two morphologically distinct types of Gamete s, and in which each individual produces only one type, a female is any individual that produces the larger type of Gamete -- called an ovum (or egg) -- and a male produces the smaller tadpole-like type -- called a sperm. This is an example of anisogamy or heterogamy, the condition wherein females and males produce Gamete s of different sizes (this is the case in humans; the human ovum is approximately 20 times larger than the human sperm cell.)
Genome	In classical genetics, the Genome of a diploid organism including eukarya refers to a full set of chromosomes or genes in a gamete; thereby, a regular somatic cell contains two full sets of Genome s. In haploid organisms, including bacteria, archaea, viruses, and mitochondria, a cell contains only a single set of the Genome usually in a single circular or contiguous linear DNA (or RNA for retroviruses.) In modern molecular biology the Genome of an organism is its hereditary information encoded in DNA (or, for retroviruses, RNA.)
Wild type	Wild type, sometimes written wildtype or wild-type, is the typical form of an organism, strain, gene, or characteristic as it occurs in nature. Wild type refers to the most common phenotype in the natural population. Wild type also refers to the allele at each locus required to produce the wild-type phenotype.
Camerae	Camerae are the spaces or chambers enclosed between two adjacent septa in the phragmocone of a nautiloid or ammonoid cephalopod. These can be seen in cross-sections of a nautilus shell and in the polished cross-sections of ammonites. In life these chambers are filled with gas, mediated by the siphuncle, and used to control buoyancy.

Fish	FISH (fluorescence in situ hybridization) is a cytogenetic technique used to detect and localize the presence or absence of specific DNA sequences on chromosomes. FISH uses fluorescent probes that bind to only those parts of the chromosome with which they show a high degree of sequence similarity. Fluorescence microscopy can be used to find out where the fluorescent probe bound to the chromosomes.
Chicken anaemia virus	Chicken anaemia virus is a virus that affects poultry. Chicken anaemia virus causes anaemia, bone marrow atrophy, and severe immunosuppression. Clinical signs of infection of Chicken anaemia virus is predominantly found in young chicks due to maternal antibodies present in most adult chickens.
Human	A Human is a member of a species of bipedal primates in the family Hominidae . DNA and fossil evidence indicates that modern Human s originated in east Africa about 200,000 years ago. When compared to other animals and primates, Human s have a highly developed brain, capable of abstract reasoning, language, introspection and problem solving.
Allele frequency	Allele frequency is the proportion of all variants (alleles) that is made up of a particular gene variant. In other words, it is the number of copies of a particular allele divided by the number of copies of all alleles at the genetic place (locus) in a population. It can be expressed for example as a percentage.
Virus	A virus is a microscopic infectious agent that can reproduce only inside a host cell. virus es infect all types of organisms: from animals and plants, to bacteria and archaea. Since the initial discovery of tobacco mosaic virus by Martinus Beijerinck in 1898, more than 5,000 types of virus have been described in detail, although most types of virus remain undiscovered.
Electrophoresis	Electrophoresis is the best-known electrokinetic phenomenon. It was discovered by Reuss in 1807. He observed that clay particles dispersed in water migrate under influence of an applied electric field.
Genetic drift	Genetic drift or allelic drift is the change in the relative frequency with which a gene variant (allele) occurs in a population that results from the fact that alleles in offspring are a random sample of those in the parents, and because of the role of chance in determining whether a given individual survives and reproduces. A population"s allele frequency is the fraction of the gene copies that share a particular form. Genetic drift is one of several evolutionary processes which lead to changes in allele frequencies over time.
History of the Polymerase Chain Reaction	The history of the Polymerase Chain Reaction has variously been described as a classic "Eureka!" moment, or as an example of cooperative teamwork between disparate researchers. A list of some of the events before, during, and after its development:

· On April 25, 1953 James D. Watson and Francis Crick publish "a radically different structure" for DNA, thereby founding the field of Molecular Genetics. Their structure involves two strands of complementary base-paired DNA, running in opposite directions as a double helix. They conclude their report saying that "It has not escaped our notice that the specific pairing we have postulated immediately suggests a possible copying mechanism for the genetic material". They are awarded the Nobel Prize in 1962.

· Starting in the mid 1950s, Arthur Kornberg begins to study the mechanism of DNA replication. By 1957 he has identified the first DNA polymerase. The enzyme is surprisingly limited, creating DNA in just one direction and requiring an existing primer to initiate copying of the template strand.

Polymerase

A Polymerase (EC 2.7.7.6/7/19/48/49) is an enzyme whose central function is associated with polymers of nucleic acids such as RNA and DNA. The primary function of a Polymerase is the polymerization of new DNA or RNA against an existing DNA or RNA template in the processes of replication and transcription. In association with a cluster of other enzymes and proteins, they take nucleotides from solvent, and catalyse the synthesis of a polynucleotide sequence against a nucleotide template strand using base-pairing interactions.

It is an accident of history that the enzymes responsible for the catalytic production of other biopolymers are not also referred to as Polymerase s.

Polymerase chain reaction

In molecular biology, the Polymerase chain reaction is a technique to amplify a single or few copies of a piece of DNA across several orders of magnitude, generating millions or more copies of a particular DNA sequence. The method relies on thermal cycling, consisting of cycles of repeated heating and cooling of the reaction for DNA melting and enzymatic replication of the DNA. Primers (short DNA fragments) containing sequences complementary to the target region along with a DNA polymerase (after which the method is named) are key components to enable selective and repeated amplification. As Polymerase chain reaction progresses, the DNA generated is itself used as a template for replication, setting in motion a chain reaction in which the DNA template is exponentially amplified.

Population

In biology, a population is the collection of inter-breeding organisms of a particular species; in sociology, a collection of human beings. Individuals within a population share a factor may be reduced by statistical means, but such a generalization may be too vague to imply anything. Demography is used extensively in marketing, which relates to economic units, such as retailers, to potential customers.

Polymorphism

Polymorphism in biophysics is the aspect of the behaviour of lipids that influences their long-range order, i.e. how they aggregate. This can be in the form of spheres of lipid molecules (micelles), pairs of layers that face one another (lamellar phase, observed in biological system as a lipid bilayer), a tubular arrangement (hexagonal), or various cubic phases (Fd3m, Im3m, Ia3m, Pn3m, and Pm3m being those discovered so far.) More complicated aggregations have also been observed, rhombohedral, tetragonal and orthorhombic phases have been observed.

CFTR

CFTR (cystic fibrosis transmembrane conductance regulator, ATP-binding cassette (sub-family C, member 7)) is a human gene that provides instructions for making a protein called the cystic fibrosis transmembrane conductance regulator. This protein functions as an ion channel across the cell membrane. Such channels are found in tissues that produce mucus, sweat, saliva, tears and digestive enzymes.

Cystic fibrosis

Cystic fibrosis is a genetic disorder known to be an inherited disease of the secretory glands, including the glands that make mucus and sweat.

The hallmarks of Cystic fibrosis are salty tasting skin, normal appetite but poor growth and poor weight gain, excess mucus production, and coughing/shortness of breath. Males can be infertile due to the condition congenital bilateral absence of the vas deferens.

Pseudomonas

Pseudomonas is a genus of gamma proteobacteria, belonging to the larger family of pseudomonads.

Recently, 16S rRNA sequence analysis has redefined the taxonomy of many bacterial species. As a result the genus Pseudomonas includes strains formerly classified in the genera Chryseomonas and Flavimonas.

Charles Robert Darwin	Charles Robert Darwin FRS (12 February 1809 - 19 April 1882) was an English naturalist who realised and presented compelling evidence that all species of life have evolved over time from common ancestors, through the process he called natural selection. The fact that evolution occurs became accepted by the scientific community and much of the general public in his lifetime, while his theory of natural selection came to be widely seen as the primary explanation of the process of evolution in the 1930s, and now forms the basis of modern evolutionary theory. In modified form, Darwin"s scientific discovery is the unifying theory of the life sciences, providing logical explanation for the diversity of life.
Population	In biology, a population is the collection of inter-breeding organisms of a particular species; in sociology, a collection of human beings. Individuals within a population share a factor may be reduced by statistical means, but such a generalization may be too vague to imply anything. Demography is used extensively in marketing, which relates to economic units, such as retailers, to potential customers.
Genetic	Genetics is the study of how living things receive common traits from previous generations. These traits are described by the Genetic information carried by a molecule called DNA. The instructions for constructing and operating an organism are contained in the organism"s DNA. Every living thing on earth has DNA in its cells. Genes are the hereditary components of DNA that occupy spots on chromosomes and determine characteristics in an organism.
Chicken anaemia virus	Chicken anaemia virus is a virus that affects poultry. Chicken anaemia virus causes anaemia, bone marrow atrophy, and severe immunosuppression. Clinical signs of infection of Chicken anaemia virus is predominantly found in young chicks due to maternal antibodies present in most adult chickens.
Gene	A Gene is the basic unit of heredity in a living organism. All living things depend on Gene s. Gene s hold the information to build and maintain their cells and pass Gene tic traits to offspring.
Genetic drift	Genetic drift or allelic drift is the change in the relative frequency with which a gene variant (allele) occurs in a population that results from the fact that alleles in offspring are a random sample of those in the parents, and because of the role of chance in determining whether a given individual survives and reproduces. A population"s allele frequency is the fraction of the gene copies that share a particular form. Genetic drift is one of several evolutionary processes which lead to changes in allele frequencies over time.
Human	A Human is a member of a species of bipedal primates in the family Hominidae . DNA and fossil evidence indicates that modern Human s originated in east Africa about 200,000 years ago. When compared to other animals and primates, Human s have a highly developed brain, capable of abstract reasoning, language, introspection and problem solving.
Leukemia inhibitory factor	Leukemia inhibitory factor an interleukin 6 class cytokine, is a chemical in cells that affects their growth and development.

Leukemia inhibitory factor derives its name from its ability to induce the terminal differentiation of myeloid leukaemic cells. Other properties attributed to the cytokine include: the growth promotion and cell differentiation of different types of target cells, influence on bone metabolism, cachexia, neural development, embryogenesis and inflammation.

Mutation

In biology, mutation s are changes to the nucleotide sequence of the genetic material of an organism. mutation s can be caused by copying errors in the genetic material during cell division, by exposure to ultraviolet or ionizing radiation, chemical mutagens, or viruses, or can be induced by the organism itself, by cellular processes such as hyper mutation . In multicellular organisms with dedicated reproductive cells, mutation s can be subdivided into germ line mutation s, which can be passed on to descendants through the reproductive cells, and somatic mutation s, which involve cells outside the dedicated reproductive group and which are not usually transmitted to descendants.

Allele

An Allele) is one of a series of different forms of a gene. The word is a short from of allelomorph , which was used in the early days of genetics to describe variant forms of a gene detected as different phenotypes. Allele s are now understood to be alternative DNA sequences at the same physical gene locus, which may or may not result in different phenotypic traits.

Drift Migration

Drift migration is the phenomenon in which migrating birds are blown off course by the winds at the time they are in flight. It is more likely to happen to birds heading south in autumn because the large numbers of inexperienced young birds are less able to compensate than the adults heading north in spring.

Drift is assisted by disorientating conditions such as mist or drizzle, and can result in large numbers of birds arriving together in an area in which they are not normally seen.

Virus

A virus is a microscopic infectious agent that can reproduce only inside a host cell. virus es infect all types of organisms: from animals and plants, to bacteria and archaea. Since the initial discovery of tobacco mosaic virus by Martinus Beijerinck in 1898, more than 5,000 types of virus have been described in detail, although most types of virus remain undiscovered.

Gene pool

In population genetics, a Gene pool is the complete set of unique alleles in a species or population.

A large Gene pool indicates extensive genetic diversity, which is associated with robust populations that can survive bouts of intense selection. Meanwhile, low genetic diversity can cause reduced biological fitness and an increased chance of extinction.

Mating

In biology, Mating is the pairing of opposite-sex or hermaphroditic organisms for copulation and, in social animals, also to raise their offspring. For animals, Mating methods include random Mating, disassortative Mating, assortative Mating, or a Mating pool.

In some birds, for example, it includes nest-building and feeding offspring.

Punnett square	The Punnett square is a diagram that is used to predict the outcome of a particular cross or breeding experiment. It is named after Reginald C. Punnett, who devised the approach, and is used by biologists to determine the probability of an offspring having a particular genotype. The Punnett square is a summary of every possible combination of one maternal allele with one paternal allele for each gene being studied in the cross.
Phenotype	A Phenotype is any observable characteristic or trait of an organism: such as its morphology, development, biochemical or physiological properties, or behavior. Phenotype s result from the expression of an organism"s genes as well as the influence of environmental factors and possible interactions between the two. The genotype of an organism is the inherited instructions it carries within its genetic code.
Blending inheritance	In Darwin"s time, biologists held to the theory of Blending inheritance--an offspring was an average of its parent. If an individual had one short parent and one tall parent, it would always be of some interim height between that of its parents. In turn, this interim height would become a theoretical extreme (either tall or short) that, in turn, bounded the potential height of its own offspring.
Appendage	An Appendage in the broadest sense is an additional or subsidiary part existing on something which can generally still function if the Appendage has never existed or is later provided or grown or natural prolongation, that protrudes from an organism"s body, such as a vertebrate"s limbs. In invertebrate biology, "Appendage" is a general term that covers any of the homologous body parts that may extend from a body segment.
Allele frequency	Allele frequency is the proportion of all variants (alleles) that is made up of a particular gene variant. In other words, it is the number of copies of a particular allele divided by the number of copies of all alleles at the genetic place (locus) in a population. It can be expressed for example as a percentage.
Dehydrogenase	A Dehydrogenase is an enzyme that oxidizes a substrate by transferring one or more hydrides (H^-) to an acceptor, usually NAD^+/$NADP^+$ or a flavin coenzyme such as FAD or FMN. · aldehyde Dehydrogenase · acetaldehyde Dehydrogenase · alcohol Dehydrogenase · glutamate Dehydrogenase (an enzyme that can convert glutamate to α-Ketoglutarate and vice versa.) · lactate Dehydrogenase · pyruvate Dehydrogenase (a common enzyme which feeds the TCA Cycle in converting Pyruvate to Acetyl CoA) · glucose-6-phosphate Dehydrogenase (involved in the pentose phosphate pathway) · glyceraldehyde-3-phosphate Dehydrogenase (involved in glycolysis)

TCA cycle examples:

· isocitrate Dehydrogenase
· alpha-ketoglutarate Dehydrogenase
· succinate Dehydrogenase
· malate Dehydrogenase .

Antibiotic resistance

Antibiotic resistance is the ability of a microorganism to withstand the effects of antibiotics. It is a specific type of drug resistance. antibiotic resistance evolves via natural selection acting upon random mutation, but it can also be engineered by applying an evolutionary stress on a population.

Horse murders

The Horse murders scandal was a form of insurance fraud in the United States in which expensive horses, many of them show jumpers, were insured against death, accident and then killed to collect the insurance money. It is not known how many horses were murdered between the mid 1970s and the mid-1990s, when a Federal Bureau of Investigation (FBI) investigation brought the horse killings to light, but the number is thought to be well over 50, and may have been as high as 100. In addition, in 1977, the heiress Helen Brach disappeared and was presumed by law enforcement agents to have been murdered by the perpetrators of these crimes, because she threatened to report their criminal activity to authorities; continuing investigations into Brach"s death began to uncover the insurance fraud in the 1990s.

Genetic variation

Genetic variation, variation in alleles of genes, occurs both within and among populations. genetic variation is important because it provides the "raw material" for natural selection.

genetic variation among individuals within a population can be identified at a variety of levels.

Amyloid

Amyloid s are insoluble fibrous protein aggregates sharing specific structural traits. Abnormal accumulation of Amyloid in organs may lead to Amyloid osis, and may play a role in various other neurodegenerative diseases.

The name Amyloid comes from the early mistaken identification of the substance as starch , based on crude iodine-staining techniques.

Chronic wasting disease

Chronic wasting disease is a transmissible spongiform encephalopathy (TSE) of mule deer, whitetailed deer, elk (wapiti), and moose. TSEs are caused by unusual infectious agents known as prions. To date, Chronic wasting disease has been found mainly in cervids (members of the deer family.)

Protein

Protein s are organic compounds made of amino acids arranged in a linear chain. The amino acids in a polymer chain are joined together by the peptide bonds between the carboxyl and amino groups of adjacent amino acid residues. The sequence of amino acids in a protein is defined by the sequence of a gene, which is encoded in the genetic code.

Buck v. Bell

Buck v. Bell, 274 U.S. 200 (1927), was the United States Supreme Court ruling that upheld a statute instituting compulsory sterilization of the mentally retarded "for the protection and health of the state." It was largely seen as an endorsement of negative eugenics--the attempt to improve the human race by eliminating "defectives" from the gene pool.

In 1924, the commonwealth of Virginia adopted a statute authorizing the compulsory sterilization of the mentally retarded for the purpose of eugenics. On September 10 of the same year, Dr. Albert Sidney Priddy, superintendent of the Virginia State Colony for Epileptics and Feebleminded, filed a petition to his Board of Directors to sterilize Carrie Buck, an 18-year-old patient at his institution whom he claimed had a mental age of 9.

Compulsory sterilization

Compulsory sterilization programs are government policies which attempt to force people to undergo surgical sterilization. In the first half of the twentieth century, many such programs were instituted in countries around the world, usually as part of eugenics programs intended to prevent the reproduction and multiplication of members of the population considered to be carriers of defective genetic traits.

Forced sterilization has been recognized as crime against humanity if the action is part of a widespread or systematic practice by the Rome Statute Explanatory Memorandum, which defines the jurisdiction of the International Criminal Court.

Alternation of phases

The Alternation of phases describes the life cycle of plants, fungi and protists. A multicellular diploid phase alternates with a multicellular haploid phase. The term can be confusing for people familiar only with the life cycle of a typical animal.

Beetles

Beetles are the group of insects with the largest number of known species. They are classified in the order Coleoptera , which contains more described species than in any other order in the animal kingdom, constituting about 25% of all known life-forms. 40% of all described insect species are Beetles , and new species are frequently discovered.

Dominance

In genetics, Dominance describes the effects of the different versions of a particular gene on the phenotype of an organism. Many animals (including humans) and plants have two copies of each gene in their genome, one inherited from each parent. The different variants of a specific gene (such as that coding for earlobes) are known as alleles.

Rat

Rat s are various medium-sized, long-tailed rodents of the superfamily Muroidea. "True Rat s" are members of the genus Rat tus, the most important of which to humans are the black Rat Rat tus Rat tus, and the brown Rat Rat tus norvegicus. Many members of other rodent genera and families are also called Rat s and share many characteristics with true Rat s.

DNA

Deoxyribonucleic acid (DNA) is a nucleic acid that contains the genetic instructions used in the development and functioning of all known living organisms and some viruses. The main role of DNA molecules is the long-term storage of information. DNA is often compared to a set of blueprints or a recipe, or a code, since it contains the instructions needed to construct other components of cells, such as proteins and RNA molecules.

Drosophila	Drosophila has long been a favorite model system for geneticists and developmental biologists studying embryogenesis. The small size, short generation time, and large brood size makes it ideal for genetic studies. Transparent embryos facilitate developmental studies.
Chromosome	A Chromosome is an organized structure of DNA and protein that is found in cells. It is a single piece of coiled DNA containing many genes, regulatory elements and other nucleotide sequences. Chromosome s also contain DNA-bound proteins, which serve to package the DNA and control its functions.
Genetic diversity	Genetic diversity is a level of biodiversity that refers to the total number of genetic characteristics in the genetic makeup of a species. It is distinguished from genetic variability, which describes the tendency of genetic characteristics to vary. The academic field of population genetics includes several hypotheses and theories regarding genetic diversity.
Gene flow	In population genetics, gene flow is the transfer of alleles of genes from one population to another. Migration into or out of a population may be responsible for a marked change in allele frequencies Immigration may also result in the addition of new genetic variants to the established gene pool of a particular species or population.
Edward	The Edward mango is a monoembryonic mango cultivar grown predominantly in Florida. It is considered by many to be among the finest tasting mangoes in the world; however, its poor yields have restrained the Edward from developing into a commercially significant variety. The Edward was first propagated in the 1920s by Edward Simmonds of the Plant Introduction Garden in Miami, Florida and is believed to be a hybrid cross of Haden and Carabao mango cultivars.
Back	The Back describes the area of horse anatomy where the saddle goes, and in popular usage extends to include the loin or lumbar region behind the thoracic vertebrae that also is crucial to a horse"s weight-carrying ability. These two sections of the vertebral column beginning at the withers, the start of the thoracic vertebrae, and extend to the last lumbar vertebra. Because horses are ridden by humans, the strength and structure of the horse"s Back is critical to the animal"s usefulness.
Cystic fibrosis	Cystic fibrosis is a genetic disorder known to be an inherited disease of the secretory glands, including the glands that make mucus and sweat. The hallmarks of Cystic fibrosis are salty tasting skin, normal appetite but poor growth and poor weight gain, excess mucus production, and coughing/shortness of breath. Males can be infertile due to the condition congenital bilateral absence of the vas deferens.
Pseudomonas	Pseudomonas is a genus of gamma proteobacteria, belonging to the larger family of pseudomonads.

Recently, 16S rRNA sequence analysis has redefined the taxonomy of many bacterial species. As a result the genus Pseudomonas includes strains formerly classified in the genera Chryseomonas and Flavimonas.

Combination

In combinatorial mathematics, a combination is an un-ordered collection of distinct elements, usually of a prescribed size and taken from a given set. (An ordered collection of distinct elements would sometimes be called a permutation, but that term is ambiguous; it can also mean "reordering of all terms", among other related notions.) Given such a set S, a combination of elements of S is just a subset of S, where, as always for (sub)sets the order of the elements is not taken into account (two lists with the same elements in different orders are considered to be the same combination.)

CFTR

CFTR (cystic fibrosis transmembrane conductance regulator, ATP-binding cassette (sub-family C, member 7)) is a human gene that provides instructions for making a protein called the cystic fibrosis transmembrane conductance regulator. This protein functions as an ion channel across the cell membrane. Such channels are found in tissues that produce mucus, sweat, saliva, tears and digestive enzymes.

Axon

An Axon or nerve fiber is a long, slender projection of a nerve cell that conducts electrical impulses away from the neuron"s cell body or soma.

An Axon is one of two types of protoplasmic protrusions that extrude from the cell body of a neuron, the other type being dendrites. Axon s are distinguished from dendrites by several features, including shape (dendrites often taper while Axon s usually maintain a constant radius), length (dendrites are restricted to a small region around the cell body while Axon s can be much longer), and function (dendrites usually receive signals while Axon s usually transmit them.)

Salmonella

Salmonella is a genus of rod-shaped, Gram-negative, non-spore forming, predominantly motile enterobacteria with diameters around 0.7 to 1.5 Âµm, lengths from 2 to 5 Âµm, and flagella which project in all directions (i.e. peritrichous.) They are chemoorganotrophs, obtaining their energy from oxidation and reduction reactions using organic sources and are facultative anaerobes; most species produce hydrogen sulfide, which can readily be detected by growing them on media containing ferrous sulfate, such as TSI. Most isolates exist in two phases; phase I is the motile phase and phase II the non-motile phase. Cultures that are non-motile upon primary culture may be swithched to the motile phase using a Craigie tube.

Typhoid fever

Typhoid fever Salmonella typhi or commonly just typhoid, is an illness caused by the bacterium Salmonella enterica serovar typhi. Common worldwide, it is transmitted by the ingestion of food or water contaminated with feces from an infected person. The bacteria then perforate through the intestinal wall and are phagocytosed by macrophages.

Lek	A Lek is a gathering of males, of certain animal species, for the purposes of competitive mating display. Leks assemble before and during the breeding season, on a daily basis. The same group of males meet at a traditional place and take up the same individual positions on an arena, each occupying and defending a small territory or court.
Genetic	Genetics is the study of how living things receive common traits from previous generations. These traits are described by the Genetic information carried by a molecule called DNA. The instructions for constructing and operating an organism are contained in the organism"s DNA. Every living thing on earth has DNA in its cells. Genes are the hereditary components of DNA that occupy spots on chromosomes and determine characteristics in an organism.
Genetic drift	Genetic drift or allelic drift is the change in the relative frequency with which a gene variant (allele) occurs in a population that results from the fact that alleles in offspring are a random sample of those in the parents, and because of the role of chance in determining whether a given individual survives and reproduces. A population"s allele frequency is the fraction of the gene copies that share a particular form. Genetic drift is one of several evolutionary processes which lead to changes in allele frequencies over time.
Mating	In biology, Mating is the pairing of opposite-sex or hermaphroditic organisms for copulation and, in social animals, also to raise their offspring. For animals, Mating methods include random Mating, disassortative Mating, assortative Mating, or a Mating pool. In some birds, for example, it includes nest-building and feeding offspring.
Mutation	In biology, mutation s are changes to the nucleotide sequence of the genetic material of an organism. mutation s can be caused by copying errors in the genetic material during cell division, by exposure to ultraviolet or ionizing radiation, chemical mutagens, or viruses, or can be induced by the organism itself, by cellular processes such as hyper mutation . In multicellular organisms with dedicated reproductive cells, mutation s can be subdivided into germ line mutation s, which can be passed on to descendants through the reproductive cells, and somatic mutation s, which involve cells outside the dedicated reproductive group and which are not usually transmitted to descendants.
Appendage	An Appendage in the broadest sense is an additional or subsidiary part existing on something which can generally still function if the Appendage has never existed or is later provided or grown or natural prolongation, that protrudes from an organism"s body, such as a vertebrate"s limbs. In invertebrate biology, "Appendage" is a general term that covers any of the homologous body parts that may extend from a body segment.

Allele	An Allele) is one of a series of different forms of a gene. The word is a short from of allelomorph , which was used in the early days of genetics to describe variant forms of a gene detected as different phenotypes. Allele s are now understood to be alternative DNA sequences at the same physical gene locus, which may or may not result in different phenotypic traits.
Drift Migration	Drift migration is the phenomenon in which migrating birds are blown off course by the winds at the time they are in flight. It is more likely to happen to birds heading south in autumn because the large numbers of inexperienced young birds are less able to compensate than the adults heading north in spring. Drift is assisted by disorientating conditions such as mist or drizzle, and can result in large numbers of birds arriving together in an area in which they are not normally seen.
Population	In biology, a population is the collection of inter-breeding organisms of a particular species; in sociology, a collection of human beings. Individuals within a population share a factor may be reduced by statistical means, but such a generalization may be too vague to imply anything. Demography is used extensively in marketing, which relates to economic units, such as retailers, to potential customers.
Snake	Snakes, like other reptiles, have a skin covered in scales. Snakes are entirely covered with scales or scutes of various shapes and sizes. Scales protect the body of the Snake, aid it in locomotion, allow moisture to be retained within, alter the surface characteristics such as roughness to aid in camouflage, and in some cases even aid in prey capture (such as Acrochordus.)
Allele frequency	Allele frequency is the proportion of all variants (alleles) that is made up of a particular gene variant. In other words, it is the number of copies of a particular allele divided by the number of copies of all alleles at the genetic place (locus) in a population. It can be expressed for example as a percentage.
Adaptation	In ocular physiology, Adaptation is the ability of the eye to adjust to various levels of darkness and light. The human eye can function from very dark to very bright levels of light -- its sensing capabilities reach across nine orders of magnitude. This means that the brightest and the darkest light signal that the eye can sense are a factor of roughly one thousand million apart.
DNA	Deoxyribonucleic acid (DNA) is a nucleic acid that contains the genetic instructions used in the development and functioning of all known living organisms and some viruses. The main role of DNA molecules is the long-term storage of information. DNA is often compared to a set of blueprints or a recipe, or a code, since it contains the instructions needed to construct other components of cells, such as proteins and RNA molecules.
DNA sequence	A DNA sequence or genetic sequence is a succession of letters representing the primary structure of a real or hypothetical DNA molecule or strand, with the capacity to carry information as described by the central dogma of molecular biology.

The possible letters are A, C, G, and T, representing the four nucleotide bases of a DNA strand -- adenine, cytosine, guanine, thymine -- covalently linked to a phosphodiester backbone. In the typical case, the sequences are printed abutting one another without gaps, as in the sequence AAAGTCTGAC, read left to right in the 5" to 3" direction.

Genetic diversity

Genetic diversity is a level of biodiversity that refers to the total number of genetic characteristics in the genetic makeup of a species. It is distinguished from genetic variability, which describes the tendency of genetic characteristics to vary.

The academic field of population genetics includes several hypotheses and theories regarding genetic diversity.

Size

Size has been one of the most interesting aspects of cephalopod science to the general public Extinct taxa are also included.

Gene

A Gene is the basic unit of heredity in a living organism. All living things depend on Gene s. Gene s hold the information to build and maintain their cells and pass Gene tic traits to offspring.

P element

A P element is a transposon that is present in the fruit fly Drosophila melanogaster and is used widely for mutagenesis and the creation of genetically modified flies used for genetic research. The P element gives rise to a phenotype known as hybrid dysgenesis

P element s seem to have first appeared in the species only in the middle of the twentieth century.

Polymorphism

Polymorphism in biophysics is the aspect of the behaviour of lipids that influences their long-range order, i.e. how they aggregate. This can be in the form of spheres of lipid molecules (micelles), pairs of layers that face one another (lamellar phase, observed in biological system as a lipid bilayer), a tubular arrangement (hexagonal), or various cubic phases (Fd3m, Im3m, Ia3m, Pn3m, and Pm3m being those discovered so far.) More complicated aggregations have also been observed, rhombohedral, tetragonal and orthorhombic phases have been observed.

Rat

Rat s are various medium-sized, long-tailed rodents of the superfamily Muroidea. "True Rat s" are members of the genus Rat tus, the most important of which to humans are the black Rat Rat tus Rat tus, and the brown Rat Rat tus norvegicus. Many members of other rodent genera and families are also called Rat s and share many characteristics with true Rat s.

Neutral theory of molecular evolution

The Neutral theory of molecular evolution is an influential theory, which was introduced with effect by Motoo Kimura in the late 1960s and early 1970s. The theory states that the vast majority of evolutionary changes at the molecular level are caused by random drift of selectively neutral mutants. Although the theory was received by some as an argument against Darwin"s theory of evolution by natural selection, Kimura maintained, and most modern evolutionary biologists agree, that the two theories are compatible: "The theory does not deny the role of natural selection in determining the course of adaptive evolution" (Kimura, 1986.)

Molecular evolution	Molecular evolution is the process of evolution at the scale of DNA, RNA, and proteins. Molecular evolution emerged as a scientific field in the 1960s as researchers from molecular biology, evolutionary biology and population genetics sought to understand recent discoveries on the structure and function of nucleic acids and protein. Some of the key topics that spurred development of the field have been the evolution of enzyme function, the use of nucleic acid divergence as a "molecular clock" to study species divergence, and the origin of non-functional or junk DNA. Recent advances in genomics, including whole-genome sequencing, high-throughput protein characterization, and bioinformatics have led to a dramatic increase in studies on the topic.
Histones	In biology, Histones are the chief protein components of chromatin. They act as spools around which DNA winds, and they play a role in gene regulation. Without Histones, the unwound DNA in chromosomes would be very long.
Protein	Protein s are organic compounds made of amino acids arranged in a linear chain. The amino acids in a polymer chain are joined together by the peptide bonds between the carboxyl and amino groups of adjacent amino acid residues. The sequence of amino acids in a protein is defined by the sequence of a gene, which is encoded in the genetic code.
David	DAVID (the Database for Annotation, Visualization and Integrated Discovery) is a free online bioinformatics resources developed by the Laboratory of Immunopathogenesis and Bioinformatics (LIB.) All tools in the DAVID Bioinformatics Resources aim to provide functional interpretation of large lists of genes derived from genomic studies, e.g. microarray and proteomics studies. DAVID can be found at http://DAVID.niaid.nih.gov or http://DAVID.abcc.ncifcrf.gov The newly updated DAVID Bioinformatics Resources consists of the DAVID Knowledgebase and five integrated, web-based functional annotation tool suites: the DAVID Gene Functional Classification Tool, the DAVID Functional Annotation Tool, the DAVID Gene ID Conversion Tool, the DAVID Gene Name Viewer and the DAVID NIAID Pathogen Genome Browser.
Jukes	The Jukes and the Kallikaks were pseudonyms for two families used as examples during the latter 19th century and early 20th century to argue that there was a genetic disposition toward anti-social behavior or low intelligence. The arguments were used to bolster advocacy of eugenics, or the pseudo-scientific breeding of human beings, by demonstrating that traits deemed socially inferior could be passed down from generation to generation. As a general concept the Jukes Family represented inherited criminality and the Kallikak Family inherited mental retardation.

Major histocompatibility complex

The Major histocompatibility complex is a large genomic region or gene family found in most vertebrates. It is the most gene-dense region of the mammalian genome and plays an important role in the immune system, autoimmunity, and reproductive success. The proteins encoded by the Major histocompatibility complex are expressed on the surface of cells in all jawed vertebrates, and display both self antigens (peptide fragments from the cell itself) and nonself antigens (e.g. fragments of invading microorganisms) to a type of white blood cell called a T cell that has the capacity to kill or co-ordinate the killing of pathogens and infected or malfunctioning cells.

Histocompatibility

Histocompatibility is the property of having the same alleles of a set of genes called the major Histocompatibility complex. These genes are expressed in most tissues as antigens, to which the immune system makes antibodies. The immune system at first makes antibodies to all sorts of antigens, including those it has never been exposed to, but stops making them to antigens present in the body.

Nick

A Nick is a discontinuity in a double stranded DNA molecule where there is no phosphodiester bond between adjacent nucleotides of one strand typically through damage or enzyme action.

Dehydrogenase

A Dehydrogenase is an enzyme that oxidizes a substrate by transferring one or more hydrides (H^-) to an acceptor, usually NAD^+/$NADP^+$ or a flavin coenzyme such as FAD or FMN.

· aldehyde Dehydrogenase
· acetaldehyde Dehydrogenase
· alcohol Dehydrogenase
· glutamate Dehydrogenase (an enzyme that can convert glutamate to α-Ketoglutarate and vice versa.)
· lactate Dehydrogenase
· pyruvate Dehydrogenase (a common enzyme which feeds the TCA Cycle in converting Pyruvate to Acetyl CoA)
· glucose-6-phosphate Dehydrogenase (involved in the pentose phosphate pathway)
· glyceraldehyde-3-phosphate Dehydrogenase (involved in glycolysis)

TCA cycle examples:

· isocitrate Dehydrogenase
· alpha-ketoglutarate Dehydrogenase
· succinate Dehydrogenase
· malate Dehydrogenase .

Escherichia

Escherichia is a genus of Gram-negative, non-spore forming, facultatively anaerobic, rod-shaped bacteria from the family Enterobacteriaceae. Inhabitants of the gastrointestinal tracts of warm-blooded animals, Escherichia species provide a portion of the microbially-derived vitamin K for their host.

	While many Escherichia are harmless commensals, particular strains of some species are human pathogens, and are known as the most common cause of urinary tract infections, significant sources of gastrointestinal disease, ranging from simple diarrhea to dysentery-like conditions, as well as a wide-range of other pathogenic states.
Escherichia coli	Escherichia coli , is a Gram negative bacterium that is commonly found in the lower intestine of warm-blooded organisms . Most E. coli strains are harmless, but some, such as serotype O157:H7, can cause serious food poisoning in humans, and are occasionally responsible for costly product recalls. The harmless strains are part of the normal flora of the gut, and can benefit their hosts by producing vitamin K_2, or by preventing the establishment of pathogenic bacteria within the intestine.
Saccharomyces	Saccharomyces is a genus in the kingdom of fungi that includes many species of yeast. Saccharomyces is from Latin meaning sugar fungi. Many members of this genus are considered very important in food production.
Saccharomyces cerevisiae	Saccharomyces cerevisiae is a species of budding yeast. It is perhaps the most useful yeast owing to its use since ancient times in baking and brewing. It is believed that it was originally isolated from the skins of grapes
Berry	The botanical definition of a Berry is a simple fruit produced from a single ovary, such as a grape or a tomato. The Berry is the most common type of fleshy fruit in which the entire ovary wall ripens into an edible pericarp. The flowers of these plants have a superior ovary formed by the fusion of two or more carpels.
Antibiotic resistance	Antibiotic resistance is the ability of a microorganism to withstand the effects of antibiotics. It is a specific type of drug resistance. antibiotic resistance evolves via natural selection acting upon random mutation, but it can also be engineered by applying an evolutionary stress on a population.
Drosophila	Drosophila has long been a favorite model system for geneticists and developmental biologists studying embryogenesis. The small size, short generation time, and large brood size makes it ideal for genetic studies. Transparent embryos facilitate developmental studies.
Chromosome	A Chromosome is an organized structure of DNA and protein that is found in cells. It is a single piece of coiled DNA containing many genes, regulatory elements and other nucleotide sequences. Chromosome s also contain DNA-bound proteins, which serve to package the DNA and control its functions.

Crow

The true Crow s are large passerine birds that form the genus Corvus in the family Corvidae. Ranging in size from the relatively small pigeon-sized jackdaws (Eurasian and Daurian) to the Common Raven of the Holarctic region and Thick-billed Raven of the highlands of Ethiopia, the 40 or so members of this genus occur on all temperate continents (except South America) and several offshore and oceanic islands (including Hawaii.) In the United States, the word Crow is used to refer to the American Crow

Human

A Human is a member of a species of bipedal primates in the family Hominidae . DNA and fossil evidence indicates that modern Human s originated in east Africa about 200,000 years ago. When compared to other animals and primates, Human s have a highly developed brain, capable of abstract reasoning, language, introspection and problem solving.

Inbreeding depression

Inbreeding depression is reduced fitness in a given population as a result of breeding of related individuals. Breeding between closely related individuals, called inbreeding, results in more recessive deleterious traits manifesting themselves. The more closely related the breeding pair is, the more homozygous deleterious genes the offspring may have, resulting in very unfit individuals.

Family

In biological classification, Family is

· a taxonomic rank. Other well-known ranks are life, domain, kingdom, phylum, class, order, genus, and species, with Family fitting between order and genus. As for the other well-known ranks, there is the option of an immediately lower rank, indicated by the prefix sub-: subfamily .
· a taxonomic unit, a taxon, in that rank. In that case the plural is families

Example: "Walnuts and Hickories belong to the Walnut Family.

What does and does not belong to each Family is determined by a taxonomist. Similarly for the question if a particular Family should be recognized at all. Often there is no exact agreement, with different taxonomists each taking a different position.

Maternal effect

A Maternal effect is a situation where the phenotype of an organism is determined not only by the environment it experiences and its genotype, but also by the environment and phenotype of its mother. In genetics, Maternal effect s occur when an organism shows the phenotype expected from the genotype of the mother, irrespective of its own genotype, often due the mother supplying mRNA or proteins to the egg. Maternal effect s can also be caused by the maternal environment independent of genotype, sometimes controlling the size, sex, or behaviour of the offspring.

Conservation genetics	Conservation genetics is an interdisciplinary science that aims to apply genetic methods to the conservation and restoration of biodiversity. Researchers involved in Conservation genetics come from a variety of fields including population genetics, molecular ecology, biology, evolutionary biology, and systematics. Genetic diversity is one of the three fundamental levels of biodiversity, so is directly important in conservation of biodiversity, though genetic factors are also important in the conservation of species and ecosystem diversity.

Chicken anaemia virus	Chicken anaemia virus is a virus that affects poultry. Chicken anaemia virus causes anaemia, bone marrow atrophy, and severe immunosuppression. Clinical signs of infection of Chicken anaemia virus is predominantly found in young chicks due to maternal antibodies present in most adult chickens.
Chromosome	A Chromosome is an organized structure of DNA and protein that is found in cells. It is a single piece of coiled DNA containing many genes, regulatory elements and other nucleotide sequences. Chromosome s also contain DNA-bound proteins, which serve to package the DNA and control its functions.
Chromosomal crossover	Chromosomal crossover is the exchange of genetic material between homologous chromosomes during prophase 1 of meiosis. This often occurs during prophase 1 of meiosis in a process called synapsis. Synapsis begins before the synaptonemal complex develops, and is not completed until near the end of prophase 1.
Population	In biology, a population is the collection of inter-breeding organisms of a particular species; in sociology, a collection of human beings. Individuals within a population share a factor may be reduced by statistical means, but such a generalization may be too vague to imply anything. Demography is used extensively in marketing, which relates to economic units, such as retailers, to potential customers.
Genetic	Genetics is the study of how living things receive common traits from previous generations. These traits are described by the Genetic information carried by a molecule called DNA. The instructions for constructing and operating an organism are contained in the organism"s DNA. Every living thing on earth has DNA in its cells. Genes are the hereditary components of DNA that occupy spots on chromosomes and determine characteristics in an organism.
Genetic drift	Genetic drift or allelic drift is the change in the relative frequency with which a gene variant (allele) occurs in a population that results from the fact that alleles in offspring are a random sample of those in the parents, and because of the role of chance in determining whether a given individual survives and reproduces. A population"s allele frequency is the fraction of the gene copies that share a particular form. Genetic drift is one of several evolutionary processes which lead to changes in allele frequencies over time.
Genetic recombination	Genetic recombination is the process by which a strand of genetic material (usually DNA; but can also be RNA) is broken and then joined to a different DNA molecule. In eukaryotes recombination commonly occurs during meiosis as chromosomal crossover between paired chromosomes. This process leads to offspring having different combinations of genes from their parents and can produce new chimeric alleles.

Genome	In classical genetics, the Genome of a diploid organism including eukarya refers to a full set of chromosomes or genes in a gamete; thereby, a regular somatic cell contains two full sets of Genome s. In haploid organisms, including bacteria, archaea, viruses, and mitochondria, a cell contains only a single set of the Genome usually in a single circular or contiguous linear DNA (or RNA for retroviruses.) In modern molecular biology the Genome of an organism is its hereditary information encoded in DNA (or, for retroviruses, RNA.)
Human	A Human is a member of a species of bipedal primates in the family Hominidae . DNA and fossil evidence indicates that modern Human s originated in east Africa about 200,000 years ago. When compared to other animals and primates, Human s have a highly developed brain, capable of abstract reasoning, language, introspection and problem solving.
Human genome	The Human genome is the genome of Homo sapiens, which is stored on 23 chromosome pairs. Twenty-two of these are autosomal chromosome pairs, while the remaining pair is sex-determining. The haploid Human genome occupies a total of just over 3 billion DNA base pairs.
Human leukocyte antigen	The Human leukocyte antigen system is the name of the major histocompatibility complex (MHC) in humans. The superlocus contains a large number of genes related to immune system function in humans. This group of genes resides on chromosome 6, and encode cell-surface antigen-presenting proteins and many other genes.
Arabidopsis thaliana	Arabidopsis thaliana , is a small flowering plant native to Europe, Asia, and northwestern Africa. A spring annual with a relatively short life cycle, Arabidopsis is popular as a model organism in plant biology and genetics. Its genome is one of the smallest plant genomes and was the first plant genome to be sequenced.
Gene	A Gene is the basic unit of heredity in a living organism. All living things depend on Gene s. Gene s hold the information to build and maintain their cells and pass Gene tic traits to offspring.
Horse murders	The Horse murders scandal was a form of insurance fraud in the United States in which expensive horses, many of them show jumpers, were insured against death, accident and then killed to collect the insurance money. It is not known how many horses were murdered between the mid 1970s and the mid-1990s, when a Federal Bureau of Investigation (FBI) investigation brought the horse killings to light, but the number is thought to be well over 50, and may have been as high as 100. In addition, in 1977, the heiress Helen Brach disappeared and was presumed by law enforcement agents to have been murdered by the perpetrators of these crimes, because she threatened to report their criminal activity to authorities; continuing investigations into Brach"s death began to uncover the insurance fraud in the 1990s.

Mustard	Mustard is a thick yellowish-brown paste with a sharp taste made from the ground seeds of a Mustard plant (white or yellow Mustard, Sinapis hirta; brown or Indian Mustard, Brassica juncea; or black Mustard, Brassica nigra.) The ground Mustard seeds are mixed with water, vinegar or other liquids, and sometimes other flavorings and spices. A strong Mustard can cause the eyes to water, sting the palate and inflame the nasal passages.
Locus	In the fields of genetics and evolutionary computation, a Locus is a fixed position on a chromosome such as the position of a biomarker that may be occupied by one or more genes. A variant of the DNA sequence at a given Locus is called an allele. The ordered list of loci known for a particular genome is called a genetic map.
Mutation	In biology, mutation s are changes to the nucleotide sequence of the genetic material of an organism. mutation s can be caused by copying errors in the genetic material during cell division, by exposure to ultraviolet or ionizing radiation, chemical mutagens, or viruses, or can be induced by the organism itself, by cellular processes such as hyper mutation . In multicellular organisms with dedicated reproductive cells, mutation s can be subdivided into germ line mutation s, which can be passed on to descendants through the reproductive cells, and somatic mutation s, which involve cells outside the dedicated reproductive group and which are not usually transmitted to descendants.
Factor XI	Factor XI or plasma thromboplastin antecedent is the zymogen form of Factor XI a, one of the enzymes of the coagulation cascade. Like many other coagulation factors, it is a serine protease. In humans, Factor XI is encoded by the F11 gene.
Fanconi anemia	Fanconi anemia is a genetic disease that affects children and adults from all ethnic backgrounds. The disease is named after the Swiss pediatrician who originally described this disorder, Guido Fanconi. It should not be confused with Fanconi syndrome, a kidney disorder also named after Fanconi.
Bubonic plague	Bubonic plague is the best known manifestation of the bacterial disease plague, caused by the bacterium Yersinia pestis (formerly known as Pasteurella pestis.) The term "Bubonic plague" was often used synonymously for plague, but it does in fact refer specifically to an infection that enters through the skin and travels through the lymphatics, as is often seen in flea-borne infections. Bubonic plague kills about 50% of infected patients in 3-7 days without treatment, and is believed by many to be the Black Death that swept through Europe in the 1340s, killing millions.
David	DAVID (the Database for Annotation, Visualization and Integrated Discovery) is a free online bioinformatics resources developed by the Laboratory of Immunopathogenesis and Bioinformatics (LIB.) All tools in the DAVID Bioinformatics Resources aim to provide functional interpretation of large lists of genes derived from genomic studies, e.g. microarray and proteomics studies. DAVID can be found at http://DAVID.niaid.nih.gov or http://DAVID.abcc.ncifcrf.gov

The newly updated DAVID Bioinformatics Resources consists of the DAVID Knowledgebase and five integrated, web-based functional annotation tool suites: the DAVID Gene Functional Classification Tool, the DAVID Functional Annotation Tool, the DAVID Gene ID Conversion Tool, the DAVID Gene Name Viewer and the DAVID NIAID Pathogen Genome Browser.

Glucose-6-phosphate dehydrogenase deficiency

Glucose-6-phosphate dehydrogenase deficiency is an X-linked recessive hereditary disease characterised by abnormally low levels of glucose-6-phosphate dehydrogenase , a metabolic enzyme involved in the pentose phosphate pathway, especially important in red blood cell metabolism. Individuals with the disease may exhibit nonimmune hemolytic anemia in response to a number of causes, most commonly infection or exposure to certain medications or chemicals. G6PD deficiency is closely linked to favism, a disorder characterized by a hemolytic reaction to consumption of broad beans, with a name derived from the Italian name of the broad bean (fava.)

Plague

Plague is a deadly infectious disease caused by the enterobacteria Yersinia pestis (Pasteurella pestis.) Plague is a zoonotic, primarily carried by rodents (most notably rats) and spread to humans via fleas. Plague is notorious throughout history, due to the unprecedented scale of death and devastation it brought.

Allele

An Allele) is one of a series of different forms of a gene. The word is a short from of allelomorph , which was used in the early days of genetics to describe variant forms of a gene detected as different phenotypes. Allele s are now understood to be alternative DNA sequences at the same physical gene locus, which may or may not result in different phenotypic traits.

Dehydrogenase

A Dehydrogenase is an enzyme that oxidizes a substrate by transferring one or more hydrides (H^-) to an acceptor, usually NAD^+/$NADP^+$ or a flavin coenzyme such as FAD or FMN.

· aldehyde Dehydrogenase
· acetaldehyde Dehydrogenase
· alcohol Dehydrogenase
· glutamate Dehydrogenase (an enzyme that can convert glutamate to α-Ketoglutarate and vice versa.)
· lactate Dehydrogenase
· pyruvate Dehydrogenase (a common enzyme which feeds the TCA Cycle in converting Pyruvate to Acetyl CoA)
· glucose-6-phosphate Dehydrogenase (involved in the pentose phosphate pathway)
· glyceraldehyde-3-phosphate Dehydrogenase (involved in glycolysis)

TCA cycle examples:

· isocitrate Dehydrogenase
· alpha-ketoglutarate Dehydrogenase
· succinate Dehydrogenase
· malate Dehydrogenase .

Anemia

Species of the genus Anemia are sometimes called flowering ferns, but this term is more commonly applied to ferns of the genus Osmunda. It is sometimes classified in family Schizaeaceae.

· A. adiantifolia - pineland fern
· A. cicutaria - hemlock fern
· A. coriacia
· A. cuneata
· A. donnell-smithii
· A. hirsuta - hairy flowering fern
· A. hirta - streambank flowering fern
· A. mexicana - Mexican flowering fern
· A. phyllitidis
· A. portoricensis - Puerto Rican flowering fern
· A. underwoodiana
· A. wrightii - Wright''s flowering fern .

Sugar

Sugar is a class of edible crystalline substances, mainly sucrose, lactose, and fructose. Human taste buds interpret its flavor as sweet. Sugar as a basic food carbohydrate primarily comes from Sugar cane and from Sugar beet, but also appears in fruit, honey, sorghum, Sugar maple (in maple syrup), and in many other sources.

Hydra

Hydra is a genus of simple fresh-water animal possessing radial symmetry. Hydras are predatory animals belonging to the phylum Cnidaria and the class Hydrozoa. They can be found in most unpolluted freshwater ponds, lakes and streams in the temperate and tropical regions by gently sweeping a collecting net through weedy areas.

Volvox

Volvox is one of the best-known chlorophytes and is the most developed in a series of genera that form spherical colonies. Each mature Volvox colony is composed of numerous flagellate cells similar to Chlamydomonas, up to 50,000 in total, and embedded in the surface of a hollow sphere or coenobium containing an extracellular matrix made of a gelatinous glycoprotein. The cells swim in a coordinated fashion, with a distinct anterior and posterior poles.

Beetles

Beetles are the group of insects with the largest number of known species. They are classified in the order Coleoptera , which contains more described species than in any other order in the animal kingdom, constituting about 25% of all known life-forms. 40% of all described insect species are Beetles , and new species are frequently discovered.

Meiosis

In biology, Meiosis is a process of reductional division in which the number of chromosomes per cell is halved. In animals, Meiosis always results in the formation of gametes, while in other organisms it can give rise to spores. As with mitosis, before Meiosis begins, the DNA in the original cell is replicated during S-phase of the cell cycle.

Artificial selection

Artificial selection describes intentional breeding for certain traits, or combination of traits. It was defined by Charles Darwin in contrast to natural selection, in which the differential reproduction of organisms with certain traits is attributed to improved survival or reproductive ability ("Darwinian fitness") and the creation of new organisms, us humans breed them for specific traits. Artificial selection can also be unintentional; it is thought that domestication of crops by early humans was largely unintentional.

Back

The Back describes the area of horse anatomy where the saddle goes, and in popular usage extends to include the loin or lumbar region behind the thoracic vertebrae that also is crucial to a horse"s weight-carrying ability. These two sections of the vertebral column beginning at the withers, the start of the thoracic vertebrae, and extend to the last lumbar vertebra. Because horses are ridden by humans, the strength and structure of the horse"s Back is critical to the animal"s usefulness.

RNA

Ribonucleic acid (RNA) is a biologically important type of molecule that consists of a long chain of nucleotide units. Each nucleotide consists of a nitrogenous base, a ribose sugar, and a phosphate. RNA is very similar to DNA, but differs in a few important structural details: in the cell, RNA is usually single-stranded, while DNA is usually double-stranded; RNA nucleotides contain ribose while DNA contains deoxyribose (a type of ribose that lacks one oxygen atom); and RNA has the base uracil rather than thymine that is present in DNA.

RNA is transcribed from DNA by enzymes called RNA polymerases and is generally further processed by other enzymes.

Ribosomal RNA

Ribosomal RNA is the central component of the ribosome, the protein manufacturing machinery of all living cells. The function of the rRNA is to provide a mechanism for decoding mRNA into amino acids and to interact with the tRNAs during translation by providing peptidyl transferase activity.The tRNA then brings the necessary amino acids corresponding to the appropriate mRNA codon.

The ribosome is composed of two subunits P, and E.

· The A site in the ribosome binds to an aminoacyl-tRNA (a tRNA bound to an amino acid.)
· The amino (NH2) group of the aminoacyl-tRNA, which contains the new amino acid, attacks the ester linkage of peptidyl-tRNA (contained within the P site), which contains the last amino acid of the growing chain, forming a new peptide bond. This reaction is catalyzed by peptidyl transferase.
· The tRNA that was holding on the last amino acid is moved to the E site, and what used to be the aminoacyl-tRNA is now the peptidyl-tRNA.

A single mRNA can be translated simultaneously by multiple ribosomes.

Both prokaryotic and eukaryotic can be broken down into two subunits (the S in 16S represents Svedberg units):

Note that the S units of the subunits cannot simply be added because they represent measures of sedimentation rate rather than of mass.

Salmonella	Salmonella is a genus of rod-shaped, Gram-negative, non-spore forming, predominantly motile enterobacteria with diameters around 0.7 to 1.5 Âµm, lengths from 2 to 5 Âµm, and flagella which project in all directions (i.e. peritrichous.) They are chemoorganotrophs, obtaining their energy from oxidation and reduction reactions using organic sources and are facultative anaerobes; most species produce hydrogen sulfide, which can readily be detected by growing them on media containing ferrous sulfate, such as TSI. Most isolates exist in two phases; phase I is the motile phase and phase II the non-motile phase. Cultures that are non-motile upon primary culture may be swithched to the motile phase using a Craigie tube.
Antibiotic resistance	Antibiotic resistance is the ability of a microorganism to withstand the effects of antibiotics. It is a specific type of drug resistance. antibiotic resistance evolves via natural selection acting upon random mutation, but it can also be engineered by applying an evolutionary stress on a population.
Rac	Rac is subfamily of the Rho family of GTPases, small (~21 kDa) signaling G proteins (more specifically a GTPases.) The subgroup include: · Rac1 · Rac2 · Rac3 · RhoG .

Trait	A trait is a distinct variant of a phenotypic character of an organism that may be inherited, environmentally determined or somewhere in between. For example, eye color is a character or abstraction of an attribute, while blue, brown and hazel are traits. A trait may be any single feature or quantifiable measurement of an organism.
Genetic	Genetics is the study of how living things receive common traits from previous generations. These traits are described by the Genetic information carried by a molecule called DNA. The instructions for constructing and operating an organism are contained in the organism"s DNA. Every living thing on earth has DNA in its cells. Genes are the hereditary components of DNA that occupy spots on chromosomes and determine characteristics in an organism.
Edward	The Edward mango is a monoembryonic mango cultivar grown predominantly in Florida. It is considered by many to be among the finest tasting mangoes in the world; however, its poor yields have restrained the Edward from developing into a commercially significant variety. The Edward was first propagated in the 1920s by Edward Simmonds of the Plant Introduction Garden in Miami, Florida and is believed to be a hybrid cross of Haden and Carabao mango cultivars.
Human	A Human is a member of a species of bipedal primates in the family Hominidae . DNA and fossil evidence indicates that modern Human s originated in east Africa about 200,000 years ago. When compared to other animals and primates, Human s have a highly developed brain, capable of abstract reasoning, language, introspection and problem solving.
Quantitative trait loci	Inheritance of quantitative traits or polygenic inheritance refers to the inheritance of a phenotypic characteristic that varies in degree and can be attributed to the interactions between two or more genes and their environment. Though not necessarily genes themselves, quantitative trait loci (quantitative trait loci s) are stretches of DNA that are closely linked to the genes that underlie the trait in question. quantitative trait loci s can be molecularly identified (for example, with PCR or AFLP) to help map regions of the genome that contain genes involved in specifying a quantitative trait.
Flower	A Flower sometimes known as a bloom or blossom, is the reproductive structure found in Flower ing plants The biological function of a Flower is to mediate the union of male sperm with female ovum in order to produce seeds. The process begins with pollination, is followed by fertilization, leading to the formation and dispersal of the seeds.
Mutation	In biology, mutation s are changes to the nucleotide sequence of the genetic material of an organism. mutation s can be caused by copying errors in the genetic material during cell division, by exposure to ultraviolet or ionizing radiation, chemical mutagens, or viruses, or can be induced by the organism itself, by cellular processes such as hyper mutation . In multicellular organisms with dedicated reproductive cells, mutation s can be subdivided into germ line mutation s, which can be passed on to descendants through the reproductive cells, and somatic mutation s, which involve cells outside the dedicated reproductive group and which are not usually transmitted to descendants.

Allele	An Allele) is one of a series of different forms of a gene. The word is a short from of allelomorph , which was used in the early days of genetics to describe variant forms of a gene detected as different phenotypes. Allele s are now understood to be alternative DNA sequences at the same physical gene locus, which may or may not result in different phenotypic traits.
Horse murders	The Horse murders scandal was a form of insurance fraud in the United States in which expensive horses, many of them show jumpers, were insured against death, accident and then killed to collect the insurance money. It is not known how many horses were murdered between the mid 1970s and the mid-1990s, when a Federal Bureau of Investigation (FBI) investigation brought the horse killings to light, but the number is thought to be well over 50, and may have been as high as 100. In addition, in 1977, the heiress Helen Brach disappeared and was presumed by law enforcement agents to have been murdered by the perpetrators of these crimes, because she threatened to report their criminal activity to authorities; continuing investigations into Brach"s death began to uncover the insurance fraud in the 1990s.
Gene	A Gene is the basic unit of heredity in a living organism. All living things depend on Gene s. Gene s hold the information to build and maintain their cells and pass Gene tic traits to offspring.
Heritability	In genetics, heritability is the proportion of phenotypic variation in a population that is attributable to genetic variation among individuals. Variation among individuals may be due to genetic and/or environmental factors. heritability analyses estimate the relative contributions of differences in genetic and non-genetic factors to the total phenotypic variance in a population.
Genetic variation	Genetic variation, variation in alleles of genes, occurs both within and among populations. genetic variation is important because it provides the "raw material" for natural selection. genetic variation among individuals within a population can be identified at a variety of levels.
Hollywood	A Hollywood is a drug slang term referring to an unusually large and often dangerous single dosage of cocaine. Depending on a user"s experience with and tolerance of the drug, a typical line measures between 50-75mg of cocaine. However, a Hollywood line can exceed 200 mg of cocaine in a single dosage.
Dominance	In genetics, Dominance describes the effects of the different versions of a particular gene on the phenotype of an organism. Many animals (including humans) and plants have two copies of each gene in their genome, one inherited from each parent. The different variants of a specific gene (such as that coding for earlobes) are known as alleles.
Twins	Twins are two offspring resulting from the same pregnancy, usually born in close succession. They can be the same or different sex. Twins can either be monozygotic (MZ, colloquially "identical") or dizygotic (DZ, colloquially "fraternal" or "non-identical".)

Adaptation	In ocular physiology, Adaptation is the ability of the eye to adjust to various levels of darkness and light. The human eye can function from very dark to very bright levels of light -- its sensing capabilities reach across nine orders of magnitude. This means that the brightest and the darkest light signal that the eye can sense are a factor of roughly one thousand million apart.
Antibiotic resistance	Antibiotic resistance is the ability of a microorganism to withstand the effects of antibiotics. It is a specific type of drug resistance. antibiotic resistance evolves via natural selection acting upon random mutation, but it can also be engineered by applying an evolutionary stress on a population.
Alpine	In the nineteenth and early twentieth centuries, many Western anthropologists classified humans into a variety of races and subraces. Of these, the name Alpines was given to a physical type of the Caucasian race predominant in central/Eastern Europe and parts of Western/Central Asia, somewhat shorter, narrower shouldered and darker skinned than those they classified as Nordics. This model was first clearly defined in William Z. Ripley"s book The Races of Europe (1899), which proposed three European categories: Teutonic (later termed Nordic), Mediterranean and Alpine.
Bumblebee	A Bumblebee (or bumble bee) is any member of the bee genus Bombus, in the family Apidae. There are over 250 known species, existing primarily in the Northern Hemisphere. Bumblebee s are social insects that are characterized by black and yellow body hairs, often in bands.
Size	Size has been one of the most interesting aspects of cephalopod science to the general public Extinct taxa are also included.
Rosemary	Rosemary is a woody, perennial herb with fragrant evergreen needle-like leaves. It is native to the Mediterranean region. It is a member of the mint family Lamiaceae, which also includes many other herbs.
Phenotype	A Phenotype is any observable characteristic or trait of an organism: such as its morphology, development, biochemical or physiological properties, or behavior. Phenotype s result from the expression of an organism"s genes as well as the influence of environmental factors and possible interactions between the two. The genotype of an organism is the inherited instructions it carries within its genetic code.
Blending inheritance	In Darwin"s time, biologists held to the theory of Blending inheritance--an offspring was an average of its parent. If an individual had one short parent and one tall parent, it would always be of some interim height between that of its parents. In turn, this interim height would become a theoretical extreme (either tall or short) that, in turn, bounded the potential height of its own offspring.

Snake	Snakes, like other reptiles, have a skin covered in scales. Snakes are entirely covered with scales or scutes of various shapes and sizes. Scales protect the body of the Snake, aid it in locomotion, allow moisture to be retained within, alter the surface characteristics such as roughness to aid in camouflage, and in some cases even aid in prey capture (such as Acrochordus.)
Wasp	A Wasp is a predatory, flying, stinging insect, with a stinger and membranous forewings and hindwings. It is related to ants and bees, with all of them being members of order Hymenoptera, but is separated from ants and bees by having a stinger and no hair; bees have hair. A rough definition of the term Wasp is any member of the aculeate family Vespidae.
Escherichia	Escherichia is a genus of Gram-negative, non-spore forming, facultatively anaerobic, rod-shaped bacteria from the family Enterobacteriaceae. Inhabitants of the gastrointestinal tracts of warm-blooded animals, Escherichia species provide a portion of the microbially-derived vitamin K for their host. While many Escherichia are harmless commensals, particular strains of some species are human pathogens, and are known as the most common cause of urinary tract infections, significant sources of gastrointestinal disease, ranging from simple diarrhea to dysentery-like conditions, as well as a wide-range of other pathogenic states.
Escherichia coli	Escherichia coli , is a Gram negative bacterium that is commonly found in the lower intestine of warm-blooded organisms . Most E. coli strains are harmless, but some, such as serotype O157:H7, can cause serious food poisoning in humans, and are occasionally responsible for costly product recalls. The harmless strains are part of the normal flora of the gut, and can benefit their hosts by producing vitamin K_2, or by preventing the establishment of pathogenic bacteria within the intestine.
Heritability of IQ	Study of the Heritability of IQ is a controversial field of research that includes biology, psychology, philosophy, sociology, and anthropology. Heritability is a measure of the relative contribution of genes to the variation of a phenotype on a given group in a specific environment. The debate about IQ heritability touches on the nature versus nurture divide, and there has been no consensus in the academic community about it even since research began in the 19th century.
Beak	The Beak bill or rostrum is an external anatomical structure of birds which is used for eating and for grooming, manipulating objects, killing prey, probing for food, courtship and feeding young. The term also refers to a similar mouthpart in some monotremes, cephalopods, cetaceans, pufferfishes, turtles, Anuran tadpoles and sirens. Comparison of bird Beak s, displaying different shapes adapted to different feeding methods.

Adaptation	In ocular physiology, Adaptation is the ability of the eye to adjust to various levels of darkness and light. The human eye can function from very dark to very bright levels of light -- its sensing capabilities reach across nine orders of magnitude. This means that the brightest and the darkest light signal that the eye can sense are a factor of roughly one thousand million apart.
Lice	Lice, are an order of over 3,000 species of wingless insects; three of which are classified as human disease agents. They are obligate ectoparasites of every avian and most mammalian orders. They are not found on Monotremes and a few eutherian orders, namely the bats, whales, dolphins and porpoises and pangolins
Rock pigeon	The Rock Pigeon is a member of the bird family Columbidae (doves and pigeons.) In common usage, this bird is often simply referred to as the "pigeon". The species includes the domestic pigeon, and escaped domestic pigeons have given rise to the feral pigeon.
Beak	The Beak bill or rostrum is an external anatomical structure of birds which is used for eating and for grooming, manipulating objects, killing prey, probing for food, courtship and feeding young. The term also refers to a similar mouthpart in some monotremes, cephalopods, cetaceans, pufferfishes, turtles, Anuran tadpoles and sirens. Comparison of bird Beak s, displaying different shapes adapted to different feeding methods.
Altruism	Altruism is a well-documented animal behaviour, which appears most obviously in kin relationships but may also be evident amongst wider social groups. In the science of ethology (the study of behavior), and more generally in the study of social evolution, Altruism refers to behavior by an individual that increases the fitness of another individual while decreasing the fitness of the actor. Research in evolutionary theory has been applied to social behaviour, including Altruism.
David	DAVID (the Database for Annotation, Visualization and Integrated Discovery) is a free online bioinformatics resources developed by the Laboratory of Immunopathogenesis and Bioinformatics (LIB.) All tools in the DAVID Bioinformatics Resources aim to provide functional interpretation of large lists of genes derived from genomic studies, e.g. microarray and proteomics studies. DAVID can be found at http://DAVID.niaid.nih.gov or http://DAVID.abcc.ncifcrf.gov The newly updated DAVID Bioinformatics Resources consists of the DAVID Knowledgebase and five integrated, web-based functional annotation tool suites: the DAVID Gene Functional Classification Tool, the DAVID Functional Annotation Tool, the DAVID Gene ID Conversion Tool, the DAVID Gene Name Viewer and the DAVID NIAID Pathogen Genome Browser.
Allele	An Allele) is one of a series of different forms of a gene. The word is a short from of allelomorph , which was used in the early days of genetics to describe variant forms of a gene detected as different phenotypes. Allele s are now understood to be alternative DNA sequences at the same physical gene locus, which may or may not result in different phenotypic traits.

Allele frequency

Allele frequency is the proportion of all variants (alleles) that is made up of a particular gene variant. In other words, it is the number of copies of a particular allele divided by the number of copies of all alleles at the genetic place (locus) in a population. It can be expressed for example as a percentage.

Spider

Spider s (order Araneae) are air-breathing chelicerate arthropods that have eight legs, and chelicerae modified into fangs that inject venom. Spider s are found world-wide on every continent except for Antarctica, and have become established in nearly every ecological niche with the exception of air and sea colonization. As of 2008, approximately 40,000 Spider species, and 109 families have been recorded by taxonomists.

Fly

Nematocera (includes Eudiptera)Brachycera

True flies are insects of the order Diptera , possessing a single pair of wings on the mesothorax and a pair of halteres, derived from the hind wings, on the metathorax.

The presence of a single pair of wings distinguishes true flies from other insects with "Fly" in their name, such as mayflies, dragonflies, damselflies, stoneflies, whiteflies, fireflies, alderflies, dobsonflies, snakeflies, sawflies, caddisflies, butterflies or scorpionflies. Some true flies have become secondarily wingless, especially in the superfamily Hippoboscoidea, or among those that are inquilines in social insect colonies.

Gene

A Gene is the basic unit of heredity in a living organism. All living things depend on Gene s. Gene s hold the information to build and maintain their cells and pass Gene tic traits to offspring.

Group

In naming cultivated plants, a Group is a formal classification category, under the International Code of Nomenclature for Cultivated Plants (ICNCP):

> ICNCP Art. 3.1: "a formal category for assembing cultivars, individual plants or assemblages of plants on the basis of defined similarity" .

The term "Group" (with a capital G) was introduced in the 2004 ICNCP, replacing the "Cultivar-Group" of the 1995 ICNCP.

A Group is united by some common trait; for example there may be a Group of yellow-flowering cultivars, a Group of cultivars with variegated leaves, a Group of cultivars resistant to a particular disease, etc.

Ectothermic

Ectothermic refers to organisms that control body temperature through external means. As a result, organisms are dependent on environmental heat sources and have relatively low metabolic rates. For example, many reptiles regulate their body temperature by basking in the sun.

Thermoregulation	Thermoregulation is the ability of an organism to keep its body temperature within certain boundaries, even when the surrounding temperature is very different. This process is one aspect of homeostasis: a dynamic state of stability between an animal"s internal environment and its external environment (the study of such processes in zoology has been called ecophysiology or physiological ecology.) If the body is unable to maintain a normal temperature and it increases significantly above normal, a condition known as heat stroke occurs.
Snake	Snakes, like other reptiles, have a skin covered in scales. Snakes are entirely covered with scales or scutes of various shapes and sizes. Scales protect the body of the Snake, aid it in locomotion, allow moisture to be retained within, alter the surface characteristics such as roughness to aid in camouflage, and in some cases even aid in prey capture (such as Acrochordus.)
Fruit	The term fruit has different meanings dependent on context, and the term is not synonymous in food preparation and biology. fruit s are the means by which flowering plants disseminate seeds, and the presence of seeds indicates that a structure is most likely a fruit though not all seeds come from fruit s. No single terminology really fits the enormous variety that is found among plant fruit s.
Size	Size has been one of the most interesting aspects of cephalopod science to the general public Extinct taxa are also included.
Virus	A virus is a microscopic infectious agent that can reproduce only inside a host cell. virus es infect all types of organisms: from animals and plants, to bacteria and archaea. Since the initial discovery of tobacco mosaic virus by Martinus Beijerinck in 1898, more than 5,000 types of virus have been described in detail, although most types of virus remain undiscovered.
Trait	A trait is a distinct variant of a phenotypic character of an organism that may be inherited, environmentally determined or somewhere in between. For example, eye color is a character or abstraction of an attribute, while blue, brown and hazel are traits. A trait may be any single feature or quantifiable measurement of an organism.
Species	In biology, a Species is: · a taxonomic rank (the basic rank of biological classification) or · a unit at that rank There are many definitions of what kind of unit a Species is (or should be.) A common definition is that of a group of organisms capable of interbreeding and producing fertile offspring, and separated from other such groups with which interbreeding does not (normally) happen. Other definitions may focus on similarity of DNA or morphology. Some Species are further subdivided into sub Species , and here also there is no close agreement on the criteria to be used.

Phenotype
A Phenotype is any observable characteristic or trait of an organism: such as its morphology, development, biochemical or physiological properties, or behavior. Phenotype s result from the expression of an organism"s genes as well as the influence of environmental factors and possible interactions between the two. The genotype of an organism is the inherited instructions it carries within its genetic code.

Flea
Flea is the common name for insects of the order Siphonaptera which are wingless insects whose mouthparts are adapted for piercing skin and sucking blood. (Some authorities use the name Aphaniptera because it is older, but names above family rank need not follow the ICZN rules of priority, so most taxonomists use the more familiar name.) Flea s are external parasites, living by hematophagy off the blood of mammals and birds.

Plants
Plants are living organisms belonging to the kingdom Plantae. They include familiar organisms such as trees, herbs, bushes, grasses, vines, ferns, mosses, and green algae. About 350,000 species of Plants, defined as seed Plants, bryophytes, ferns and fern allies, are estimated to exist currently.

Begonia
Begonia is a genus in the flowering plant family Begonia ceae. The only other member of the family Begonia ceae is Hillebrandia, a genus with a single species in the Hawaiian Islands. The genus Sym Begonia is now included in Begonia

Flower
A Flower sometimes known as a bloom or blossom, is the reproductive structure found in Flower ing plants The biological function of a Flower is to mediate the union of male sperm with female ovum in order to produce seeds. The process begins with pollination, is followed by fertilization, leading to the formation and dispersal of the seeds.

Antibiotic resistance
Antibiotic resistance is the ability of a microorganism to withstand the effects of antibiotics. It is a specific type of drug resistance. antibiotic resistance evolves via natural selection acting upon random mutation, but it can also be engineered by applying an evolutionary stress on a population.

Fuchsia
Fuchsia is a genus of flowering plants, mostly shrubs, and can grow long shoots, which were identified by Charles Plumier in the late-17th century, and named by Plumier in 1703 after the German botanist Leonhart Fuchs . The English name Fuchsia s is frequently misspelled "fuschias".

There are about 100-110 species of Fuchsia

Pollen
Pollen is a fine to coarse powder consisting of microgametophytes (Pollen grains), which produce the male gametes (sperm cells) of seed plants. A hard coat covering the Pollen grain protects the sperm cells during the process of their movement between the stamens of the flower to the pistil of the next flower. Individual Pollen grains are small enough to require magnification to see detail.

Pollen tube	The Pollen tube of most seed plants acts as a conduit to transport sperm cells from the pollen grain, either from the stigma (in flowering plants or angiosperms) to the ovules at the base of the pistil and does not convey sperm cells to the egg. Like ferns, other basal land plants, and many algae, these gymnosperms have flagellate sperm, which swim through a watery fluid to fertilize the egg cells.
Genetic	Genetics is the study of how living things receive common traits from previous generations. These traits are described by the Genetic information carried by a molecule called DNA. The instructions for constructing and operating an organism are contained in the organism"s DNA. Every living thing on earth has DNA in its cells. Genes are the hereditary components of DNA that occupy spots on chromosomes and determine characteristics in an organism.
Genetic variation	Genetic variation, variation in alleles of genes, occurs both within and among populations. genetic variation is important because it provides the "raw material" for natural selection. genetic variation among individuals within a population can be identified at a variety of levels.
Leaf	In botany, a Leaf is an above-ground plant organ specialized for photosynthesis. For this purpose, a Leaf is typically flat (laminar) and thin. There is continued debate about whether the flatness of leaves [[Natural selection | evolved] to expose the chloroplasts to more light or to increase the absorption of carbon dioxide.
Beetles	Beetles are the group of insects with the largest number of known species. They are classified in the order Coleoptera , which contains more described species than in any other order in the animal kingdom, constituting about 25% of all known life-forms. 40% of all described insect species are Beetles , and new species are frequently discovered.
Feathers	Feathers are one of the epidermal growths that form the distinctive outer covering on birds. They are considered the most complex integumentary structures found in vertebrates. They are among the outstanding characteristics that distinguish the extant Aves from other living groups.
Noah	Noah was the name of species of ox called gaur, cloned and gestated in the womb of a cow named Bessie. Gaur is a vulnerable species according to the IUCN. The baby bull gaur was delivered on January 8, 2001, but died within just 48 hours of a common dysentery on January 10, 2001. Noah"s condition was monitored by Dr. Jonathan Hill and his teammates in Iowa.
Fermentation	Fermentation is the process of deriving energy from the oxidation of organic compounds, such as carbohydrates, using an endogenous electron acceptor, which is usually an organic compound. This is in contrast to cellular respiration, where electrons are donated to an exogenous electron acceptor, such as oxygen, via an electron transport chain. Fermentation does not necessarily have to be carried out in an anaerobic environment.

Cell	The Cell is the structural and functional unit of all known living organisms. It is the smallest unit of an organism that is classified as living, and is often called the building block of life. Some organisms, such as most bacteria, are unicellular (consist of a single Cell.)
Genome	In classical genetics, the Genome of a diploid organism including eukarya refers to a full set of chromosomes or genes in a gamete; thereby, a regular somatic cell contains two full sets of Genome s. In haploid organisms, including bacteria, archaea, viruses, and mitochondria, a cell contains only a single set of the Genome usually in a single circular or contiguous linear DNA (or RNA for retroviruses.) In modern molecular biology the Genome of an organism is its hereditary information encoded in DNA (or, for retroviruses, RNA.)
Apert syndrome	Apert syndrome is a form of acrocephalosyndactyly, a congenital disorder characterized by malformations of the skull, face, hands and feet. It is classified as a branchial arch syndrome, affecting the first branchial arch, the precursor of the maxilla and mandible. Disturbances in the development of the branchial arches in fetal development create lasting and widespread effects.
Fibroblast	A Fibroblast is a type of cell that synthesizes the extracellular matrix and collagen, the structural framework (stroma) for animal tissues, and play a critical role in wound healing. They are the most common cells of connective tissue in animals. Fibroblast s and fibrocytes are two states of the same cells, the former being the activated state, the latter the less active state, concerned with maintenance.
Fibroblast growth factor	Fibroblast growth factor s are a family of growth factors involved in angiogenesis, wound healing, and embryonic development. The Fibroblast growth factor s are heparin-binding proteins and interactions with cell-surface associated heparan sulfate proteoglycans have been shown to be essential for Fibroblast growth factor signal transduction. Fibroblast growth factor s are key players in the processes of proliferation and differentiation of wide variety of cells and tissues.
Receptor	In biochemistry, a Receptor is a protein molecule, embedded in either the plasma membrane or cytoplasm of a cell, to which a mobile signaling (or "signal") molecule may attach. A molecule which binds to a Receptor is called a "ligand," and may be a peptide (such as a neurotransmitter), a hormone, a pharmaceutical drug, or a toxin, and when such binding occurs, the Receptor undergoes a conformational change which ordinarily initiates a cellular response. However, some ligands merely block Receptor s without inducing any response (e.g. antagonists.)

Mutation

In biology, mutation s are changes to the nucleotide sequence of the genetic material of an organism. mutation s can be caused by copying errors in the genetic material during cell division, by exposure to ultraviolet or ionizing radiation, chemical mutagens, or viruses, or can be induced by the organism itself, by cellular processes such as hyper mutation . In multicellular organisms with dedicated reproductive cells, mutation s can be subdivided into germ line mutation s, which can be passed on to descendants through the reproductive cells, and somatic mutation s, which involve cells outside the dedicated reproductive group and which are not usually transmitted to descendants.

Human	A Human is a member of a species of bipedal primates in the family Hominidae . DNA and fossil evidence indicates that modern Human s originated in east Africa about 200,000 years ago. When compared to other animals and primates, Human s have a highly developed brain, capable of abstract reasoning, language, introspection and problem solving.
Sexual dimorphism	Sexual dimorphism is the systematic difference in form between individuals of different sex in the same species. Examples include colour (specifically referred to as sexual dichromatism), size, and the presence or absence of parts of the body used in courtship displays or fights, such as ornamental feathers, horns, antlers or tusks. Peacock courting peahen In many species, including most mammals, the male is larger than the female.
Charles Robert Darwin	Charles Robert Darwin FRS (12 February 1809 - 19 April 1882) was an English naturalist who realised and presented compelling evidence that all species of life have evolved over time from common ancestors, through the process he called natural selection. The fact that evolution occurs became accepted by the scientific community and much of the general public in his lifetime, while his theory of natural selection came to be widely seen as the primary explanation of the process of evolution in the 1930s, and now forms the basis of modern evolutionary theory. In modified form, Darwin"s scientific discovery is the unifying theory of the life sciences, providing logical explanation for the diversity of life.
Germ	The germ of a cereal is the reproductive part that germinates to grow into a plant; it is the embryo of the seed. Along with bran, germ is often a by-product of the milling that produces refined grain products. Cereal grains and their components, such as wheat germ, rice bran, and maize may be used as a source from which vegetable oil is extracted, or used directly as a food ingredient.
Size	Size has been one of the most interesting aspects of cephalopod science to the general public Extinct taxa are also included.
Species	In biology, a Species is: · a taxonomic rank (the basic rank of biological classification) or · a unit at that rank There are many definitions of what kind of unit a Species is (or should be.) A common definition is that of a group of organisms capable of interbreeding and producing fertile offspring, and separated from other such groups with which interbreeding does not (normally) happen. Other definitions may focus on similarity of DNA or morphology. Some Species are further subdivided into sub Species , and here also there is no close agreement on the criteria to be used.

Egg	Phasmatodea, like a ghost, resembles twigs leaves and sometimes resembles lichen. The Egg is oval with 5 parts making the Egg, the capitulum-small nob on end of Egg, operculum-the ring holding the capitulum and is in between the shell and the capitulum. The shell which is the larger oval shaped part to the Egg has the nymph and the sheet separating the yolk and the shell to which the nymph emerges from the connected oper and capitulum, sometimes when the nymph emerges, say the tropidoderus childrenii when the nymph hatches with the Egg still attached to the hind legs causing leg dis-formation.
Orangutans	The Orangutans are a species of great apes. Known for their intelligence, they live in trees and are the largest living arboreal animal. They have longer arms than other great apes, and their hair is typically reddish-brown, instead of the brown or black hair typical of other great apes.
Fruit	The term fruit has different meanings dependent on context, and the term is not synonymous in food preparation and biology. fruit s are the means by which flowering plants disseminate seeds, and the presence of seeds indicates that a structure is most likely a fruit though not all seeds come from fruit s. No single terminology really fits the enormous variety that is found among plant fruit s.
Adaptation	In ocular physiology, Adaptation is the ability of the eye to adjust to various levels of darkness and light. The human eye can function from very dark to very bright levels of light -- its sensing capabilities reach across nine orders of magnitude. This means that the brightest and the darkest light signal that the eye can sense are a factor of roughly one thousand million apart.
Metabolism	Metabolism is the set of chemical reactions that occur in living organisms in order to maintain life. These processes allow organisms to grow and reproduce, maintain their structures, and respond to their environments. Metabolism is usually divided into two categories.
Rat	Rat s are various medium-sized, long-tailed rodents of the superfamily Muroidea. "True Rat s" are members of the genus Rat tus, the most important of which to humans are the black Rat Rat tus Rat tus, and the brown Rat Rat tus norvegicus. Many members of other rodent genera and families are also called Rat s and share many characteristics with true Rat s.
Mating	In biology, Mating is the pairing of opposite-sex or hermaphroditic organisms for copulation and, in social animals, also to raise their offspring. For animals, Mating methods include random Mating, disassortative Mating, assortative Mating, or a Mating pool. In some birds, for example, it includes nest-building and feeding offspring.
Ceratitis capitata	Ceratitis capitata, the Mediterranean fruit fly is a species of fruit fly capable of wreaking extensive damage to a wide range of fruit crops. It is native to the Mediterranean area, but has spread invasively to many parts of the world, including Australasia and North and South America. Adult medflies lay their eggs under the skins of fruit, particularly where the skin is already broken.

Fly

Nematocera (includes Eudiptera)Brachycera

True flies are insects of the order Diptera , possessing a single pair of wings on the mesothorax and a pair of halteres, derived from the hind wings, on the metathorax.

The presence of a single pair of wings distinguishes true flies from other insects with "Fly" in their name, such as mayflies, dragonflies, damselflies, stoneflies, whiteflies, fireflies, alderflies, dobsonflies, snakeflies, sawflies, caddisflies, butterflies or scorpionflies. Some true flies have become secondarily wingless, especially in the superfamily Hippoboscoidea, or among those that are inquilines in social insect colonies.

Frog

Frog s are amphibians in the order Anura , formerly referred to as Salientia . The name Frog derives from Old English Frog ga, , cognate with Sanskrit plava , probably deriving from Proto-Indo-European praw = "to jump".

Most Frog s are characterized by long hind legs, a short body, webbed digits (fingers or toes), protruding eyes and the absence of a tail.

Gene

A Gene is the basic unit of heredity in a living organism. All living things depend on Gene s. Gene s hold the information to build and maintain their cells and pass Gene tic traits to offspring.

Metamorphosis

Metamorphosis is a biological process by which an animal physically develops after birth or hatching, involving a conspicuous and relatively abrupt change in the animal"s form or body structure through cell growth and differentiation. Some insects, amphibians, mollusks, crustaceans, Cnidarians, echinoderms and tunicates undergo Metamorphosis, which is usually (but not always) accompanied by a change of habitat or behavior.

Scientific usage of the term is exclusive, and is not applied to general aspects of cell growth, including rapid growth spurts.

Leg

A Leg is a limb on a living thing"s body that supports the rest of the animal above the ground between the ankle and the hip and is used for locomotion. The end of the Leg farthest from the animal"s body is often either modified or attached to another structure that is modified to disperse the animal"s weight on the ground In bipedal vertebrate animals, the two lower limbs are usually referred to as the " Leg s" and the two upper limbs as the "arms" or "wings" as the case may be.

Senescence

Senescence is a process induced by evolution into an organism"s genetic make up so that it may live to its healthiest until its reproductive age and die slowly and gradually thereafter. senescence encompasses all of the biological processes of a living organism"s approaching an advanced age (i.e., the combination of processes of deterioration which follow the period of development of an organism.) The word senescence is derived from the Latin word senex, meaning "old man" or "old age" or "advanced in age".

Beetles

Beetles are the group of insects with the largest number of known species. They are classified in the order Coleoptera , which contains more described species than in any other order in the animal kingdom, constituting about 25% of all known life-forms. 40% of all described insect species are Beetles , and new species are frequently discovered.

DNA

Deoxyribonucleic acid (DNA) is a nucleic acid that contains the genetic instructions used in the development and functioning of all known living organisms and some viruses. The main role of DNA molecules is the long-term storage of information. DNA is often compared to a set of blueprints or a recipe, or a code, since it contains the instructions needed to construct other components of cells, such as proteins and RNA molecules.

Fish

FISH (fluorescence in situ hybridization) is a cytogenetic technique used to detect and localize the presence or absence of specific DNA sequences on chromosomes. FISH uses fluorescent probes that bind to only those parts of the chromosome with which they show a high degree of sequence similarity. Fluorescence microscopy can be used to find out where the fluorescent probe bound to the chromosomes.

Ivy

Hedera is a genus of 15 species of climbing or ground-creeping evergreen woody plants in the family Araliaceae, native to the Atlantic Islands, western, central and southern Europe, northwestern Africa and across central-southern Asia east to Japan. On suitable surfaces " href="/wiki/Rock_(geology)">rock faces), they are able to climb to at least 25-30 metres above the basal ground level.

Regional English common names for Ivy include Bindwood and Lovestone .

Plants

Plants are living organisms belonging to the kingdom Plantae. They include familiar organisms such as trees, herbs, bushes, grasses, vines, ferns, mosses, and green algae. About 350,000 species of Plants, defined as seed Plants, bryophytes, ferns and fern allies, are estimated to exist currently.

Radish

The Radish (Raphanus sativus) is an edible root vegetable of the Brassicaceae family that was domesticated in Europe in pre-Roman times. They are grown and consumed throughout the world. Radish es have numerous varieties, varying in size, color and duration of required cultivation time.

Dioecious

Dioecious species are whose members can produce only one type of gamete; each individual organism belonging to a Dioecious species is distinctly male or female . The majority of animal species are Dioecious. In plant sexuality, there are also Dioecious species.

Pollen

Pollen is a fine to coarse powder consisting of microgametophytes (Pollen grains), which produce the male gametes (sperm cells) of seed plants. A hard coat covering the Pollen grain protects the sperm cells during the process of their movement between the stamens of the flower to the pistil of the next flower. Individual Pollen grains are small enough to require magnification to see detail.

Nettle	Nettle is the common name for between 30-45 species of flowering plants of the genus Urtica in the family Urticaceae, with a cosmopolitan though mainly temperate distribution. They are mostly herbaceous perennial plants, but some are annual and a few are shrubby. The most prominent member of the genus is the stinging Nettle Urtica dioica, native to Europe, north Africa, Asia, and North America.

Altruism	Altruism is a well-documented animal behaviour, which appears most obviously in kin relationships but may also be evident amongst wider social groups. In the science of ethology (the study of behavior), and more generally in the study of social evolution, Altruism refers to behavior by an individual that increases the fitness of another individual while decreasing the fitness of the actor. Research in evolutionary theory has been applied to social behaviour, including Altruism.
Crow	The true Crow s are large passerine birds that form the genus Corvus in the family Corvidae. Ranging in size from the relatively small pigeon-sized jackdaws (Eurasian and Daurian) to the Common Raven of the Holarctic region and Thick-billed Raven of the highlands of Ethiopia, the 40 or so members of this genus occur on all temperate continents (except South America) and several offshore and oceanic islands (including Hawaii.) In the United States, the word Crow is used to refer to the American Crow
Charles Robert Darwin	Charles Robert Darwin FRS (12 February 1809 - 19 April 1882) was an English naturalist who realised and presented compelling evidence that all species of life have evolved over time from common ancestors, through the process he called natural selection. The fact that evolution occurs became accepted by the scientific community and much of the general public in his lifetime, while his theory of natural selection came to be widely seen as the primary explanation of the process of evolution in the 1930s, and now forms the basis of modern evolutionary theory. In modified form, Darwin"s scientific discovery is the unifying theory of the life sciences, providing logical explanation for the diversity of life.
Microsatellite	Microsatellite s are polymorphic loci present in nuclear and organellar DNA that consist of repeating units of 1-6 base pairs in length. They are typically neutral, co-dominant and are used as molecular markers which have wide-ranging applications in the field of genetics, including kinship and population studies. Microsatellite s can also be used to study gene dosage (looking for duplications or deletions of a particular genetic region.)
David	DAVID (the Database for Annotation, Visualization and Integrated Discovery) is a free online bioinformatics resources developed by the Laboratory of Immunopathogenesis and Bioinformatics (LIB.) All tools in the DAVID Bioinformatics Resources aim to provide functional interpretation of large lists of genes derived from genomic studies, e.g. microarray and proteomics studies. DAVID can be found at http://DAVID.niaid.nih.gov or http://DAVID.abcc.ncifcrf.gov The newly updated DAVID Bioinformatics Resources consists of the DAVID Knowledgebase and five integrated, web-based functional annotation tool suites: the DAVID Gene Functional Classification Tool, the DAVID Functional Annotation Tool, the DAVID Gene ID Conversion Tool, the DAVID Gene Name Viewer and the DAVID NIAID Pathogen Genome Browser.

Tiger

The Tiger is a member of the Felidae family; the largest of the four "big cats" in the genus Panthera. Native to much of eastern and southern Asia, the Tiger is an apex predator and an obligate carnivore. Reaching up to 4 metres (13 ft) in total length and weighing up to 300 kilograms (660 pounds), the larger Tiger subspecies are comparable in size to the biggest extinct felids.

Bee

Bee s are flying insects closely related to wasps and ants, and are known for their roles of producing honey and Bee swax and pollination. Bee s are a monophyletic lineage within the superfamily Apoidea, presently classified by the unranked taxon name Anthophila. There are nearly 20,000 known species of Bee s in nine recognized families, though many are undescribed and the actual number is probably higher.

Salamander

Salamander is a common name of approximately 500 species of amphibians. They are typically characterized by their slender bodies, short noses, and long tails. All known fossils and extinct species fall under the order Caudata, while sometimes the extant species are grouped together as the Urodela.

Tadpole

A Tadpole, polliwog (also pollywog, polliwig, polwig or pollywiggle (also polliwiggle, polwiggle particularly of an anuran.

The name "Tadpole" is from Middle English taddepol, made up of the elements tadde, "toad", and pol, "head" " href="/wiki/Poll_(livestock)">poll".) Similarly, "polliwog" and "pollywiggle" are from Middle English polwigle, made up of the same pol and wiglen, "to wiggle".

Chemical

A chemical substance is a material with a specific chemical composition.

A common example of a chemical substance is pure water; it has the same properties and the same ratio of hydrogen to oxygen whether it is isolated from a river or made in a laboratory. Some typical chemical substances are diamond, gold, salt (sodium chloride) and sugar (sucrose.)

Kin recognition

Kin recognition is animals" abilities to distinguish between close genetic kin and non-kin. In evolutionary biology and in psychology, such capabilities are presumed to have evolved to serve the adaptive functions of kin altruism and inbreeding avoidance. Apart from a few exceptional cases, such as the green-beard effect, Kin recognition is achieved via several cue-based mechanisms, such as imprinting and phenotype matching.

Major histocompatibility complex

The Major histocompatibility complex is a large genomic region or gene family found in most vertebrates. It is the most gene-dense region of the mammalian genome and plays an important role in the immune system, autoimmunity, and reproductive success. The proteins encoded by the Major histocompatibility complex are expressed on the surface of cells in all jawed vertebrates, and display both self antigens (peptide fragments from the cell itself) and nonself antigens (e.g. fragments of invading microorganisms) to a type of white blood cell called a T cell that has the capacity to kill or co-ordinate the killing of pathogens and infected or malfunctioning cells.

Histocompatibility

Histocompatibility is the property of having the same alleles of a set of genes called the major Histocompatibility complex. These genes are expressed in most tissues as antigens, to which the immune system makes antibodies. The immune system at first makes antibodies to all sorts of antigens, including those it has never been exposed to, but stops making them to antigens present in the body.

Sense

Sense s are the physiological methods of perception. The Sense s and their operation, classification, and theory are overlapping topics studied by a variety of fields, most notably neuroscience, cognitive psychology (or cognitive science), and philosophy of perception. The nervous system has a specific sensory system, or organ, dedicated to each Sense

Allele

An Allele) is one of a series of different forms of a gene. The word is a short from of allelomorph , which was used in the early days of genetics to describe variant forms of a gene detected as different phenotypes. Allele s are now understood to be alternative DNA sequences at the same physical gene locus, which may or may not result in different phenotypic traits.

Dictyostelium discoideum

Dictyostelium discoideum is a species of soil-living amoeba belonging to the phylum Mycetozoa. D. discoideum, commonly referred to as slime mold, is a primitive eukaryote that transitions from a collection of unicellular amoebae into a multicellular slug and then into a fruiting body within its life time. D. discoideum has a unique asexual lifecycle that consists of four stages: vegetative, aggregation, migration, and culmination.

Wild type

Wild type, sometimes written wildtype or wild-type, is the typical form of an organism, strain, gene, or characteristic as it occurs in nature. Wild type refers to the most common phenotype in the natural population. Wild type also refers to the allele at each locus required to produce the wild-type phenotype.

Cell

The Cell is the structural and functional unit of all known living organisms. It is the smallest unit of an organism that is classified as living, and is often called the building block of life. Some organisms, such as most bacteria, are unicellular (consist of a single Cell.)

Eusociality

Eusociality is a term used for the highest level of social organization in a hierarchical classification. The term "eusocial" was introduced in 1966 by Suzanne Batra and given a more definitive meaning by E. O. Wilson. It was originally defined to include those organisms that had certain features:

· Reproductive division of labor (with or without sterile castes)
· Overlapping generations
· Cooperative care of young

The lower levels of social organization, presociality, were classified using different terms, including presocial, subsocial, semisocial, parasocial and quasisocial.

The most familiar examples are insects such as ants, bees, and wasps (order Hymenoptera), as well as termites (order Isoptera), all with reproductive queens and more or less sterile workers and/or soldiers.

Allele frequency

Allele frequency is the proportion of all variants (alleles) that is made up of a particular gene variant. In other words, it is the number of copies of a particular allele divided by the number of copies of all alleles at the genetic place (locus) in a population. It can be expressed for example as a percentage.

Hox genes

Hox genes are a group of related genes that specify the anterior-posterior axis and segment identity of metazoan organisms during early embryonic development. These genes are critical for the proper number and placement of embryonic segment structures (such as legs, antennae, and eyes.)

Homeotic genes are defined by a DNA sequence known as the homeobox, which is a sequence of 180 nucleotides that code for a protein domain known as the homeodomain.

Hymenoptera

Hymenoptera is one of the larger orders of insects, comprising the sawflies, wasps, bees, and ants. The name refers to the membranous wings of the insects, and is derived from the Ancient Greek á½'μÎ®ν : membrane and πτερÏŒν (pteron): wing. The hindwings are connected to the forewings by a series of hooks called hamuli.

Plants

Plants are living organisms belonging to the kingdom Plantae. They include familiar organisms such as trees, herbs, bushes, grasses, vines, ferns, mosses, and green algae. About 350,000 species of Plants, defined as seed Plants, bryophytes, ferns and fern allies, are estimated to exist currently.

Thrips

Thrips are tiny, slender insects with fringed wings .) Other common names for Thrips include thunderflies, thunderbugs, storm flies, and corn lice. Thrips species feed on a large variety of sources both plant and animal by puncturing them and sucking up the contents.

Gene expression

Gene expression is the process by which information from a gene is used in the synthesis of a functional gene product. These products are often proteins, but in non-protein coding genes such as rRNA genes or tRNA genes, the product is a functional RNA.

Several steps in the Gene expression process may be modulated, including the transcription, RNA splicing, translation, and post-translational modification of a protein. Gene regulation gives the cell control over structure and function, and is the basis for cellular differentiation, morphogenesis and the versatility and adaptability of any organism.

Antibiotic resistance

Antibiotic resistance is the ability of a microorganism to withstand the effects of antibiotics. It is a specific type of drug resistance. antibiotic resistance evolves via natural selection acting upon random mutation, but it can also be engineered by applying an evolutionary stress on a population.

Mate	Mate is a traditional South American infusion. It is prepared from steeping dried leaves of yerba Mate in hot water. It is the national drink in Argentina, Paraguay, and Uruguay, and drinking it is a common social practice in parts of Brazil, Chile, eastern Bolivia, Lebanon, Syria, Argentina and Turkey.
Nest	A Nest is a place of refuge to hold an animal"s eggs and/or provide a place to live or raise offspring. They are usually made of some organic material such as twigs, grass, and leaves; or may simply be a depression in the ground, or a hole in a tree, rock or building. Human-made materials, such as string, plastic, cloth, hair or paper, may be used.
Human	A Human is a member of a species of bipedal primates in the family Hominidae . DNA and fossil evidence indicates that modern Human s originated in east Africa about 200,000 years ago. When compared to other animals and primates, Human s have a highly developed brain, capable of abstract reasoning, language, introspection and problem solving.
Paper wasp	Paper wasp s are 3/4 to 1 inch (1.9 to 2.5 cm)-long wasps that gather fibers from dead wood and plant stems, which they mix with saliva, and use to construct water-resistant nests made of gray or brown papery material. Paper wasp s are also sometimes called umbrella wasps, due to the distinctive design of their nests or other regional variants such as Trinidad ' Tobago"s use of Jack Spaniard. European Paper wasp (Polistes dominula) The name " Paper wasp s" typically refers to members of the vespid subfamily Polistinae, though it often colloquially includes members of the subfamilies Vespinae (hornets and yellowjackets) and Stenogastrinae, which also make nests out of paper.
Size	Size has been one of the most interesting aspects of cephalopod science to the general public Extinct taxa are also included.
Dilemma	Dilemma is a genus of marine bivalves of the family Poromyidae. The genus is remarkable for encompassing predators of isopods and ostracods, unusual for sessile molluscs. · Dilemma frumarkernorum Leal, 2008 · Dilemma inexpectatum (Crozier, 1966) · Dilemma japonicum Sasaki and Leal, 2008 · Dilemma spectralis Leal, 2008
Fruit	The term fruit has different meanings dependent on context, and the term is not synonymous in food preparation and biology. fruit s are the means by which flowering plants disseminate seeds, and the presence of seeds indicates that a structure is most likely a fruit though not all seeds come from fruit s. No single terminology really fits the enormous variety that is found among plant fruit s.

Territory

In ethology, sociobiology and behavioral ecology, the term Territory refers to any sociographical area that an animal of a particular species consistently defends against conspecifics (and, occasionally, animals of other species.) Animals that defend territories in this way are referred to as territorial.

The idea of animal territories was first introduced by the British ornithologist Eliot Howard in a book published in 1920.

Black bear	The Black Bear, also known as the North American Black Bear or the American Black Bear is the most common bear species native to North America. It lives throughout much of the continent, from northern Alaska south into Mexico and from the Atlantic Ocean to the Pacific Ocean. This includes 41 of the 50 U.S. states, all Canadian provinces except Prince Edward Island, and portions of northern Mexico.
California poppy	The California poppy is native to grassy and open areas from sea level to 2,000m (6,500 feet) altitude in the western United States throughout California, extending to Oregon, southern Washington, Nevada, Arizona, New Mexico, and in Mexico in Sonora and northwest Baja California. Botanical illustration of Eschscholzia californica It can grow 5-60 cm tall, with alternately branching glaucous blue-green foliage. The leaves are ternately divided into round, lobed segments.
Clam	In the USA, the word "Clam" can be used in several different ways: one, is as a general term covering all bivalve mollusks. The word can also be used in a more limited sense, to mean bivalves which burrow in sediment, as opposed to ones which attach themselves to the substrate (for example oysters and mussels), or ones which can swim and are migratory, like scallops. In addition the word "Clam" can be used in an even more limited sense, to mean one or more species of commonly consumed marine bivalves, as in the phrase Clam chowder, meaning a thick shellfish soup usually made using the hard Clam.
Leukemia inhibitory factor	Leukemia inhibitory factor an interleukin 6 class cytokine, is a chemical in cells that affects their growth and development. Leukemia inhibitory factor derives its name from its ability to induce the terminal differentiation of myeloid leukaemic cells. Other properties attributed to the cytokine include: the growth promotion and cell differentiation of different types of target cells, influence on bone metabolism, cachexia, neural development, embryogenesis and inflammation.
Oyster	The common name Oyster is used for a number of different groups of bivalve mollusks, most of which live in marine habitats or brackish water. The shell consists of two usually highly calcified valves which surround a soft body. Gills filter plankton from the water, and strong adductor muscles are used to hold the shell closed.
Thrips	Thrips are tiny, slender insects with fringed wings .) Other common names for Thrips include thunderflies, thunderbugs, storm flies, and corn lice. Thrips species feed on a large variety of sources both plant and animal by puncturing them and sucking up the contents.
Adaptation	In ocular physiology, Adaptation is the ability of the eye to adjust to various levels of darkness and light. The human eye can function from very dark to very bright levels of light -- its sensing capabilities reach across nine orders of magnitude. This means that the brightest and the darkest light signal that the eye can sense are a factor of roughly one thousand million apart.

Egg	Phasmatodea, like a ghost, resembles twigs leaves and sometimes resembles lichen. The Egg is oval with 5 parts making the Egg, the capitulum-small nob on end of Egg, operculum-the ring holding the capitulum and is in between the shell and the capitulum. The shell which is the larger oval shaped part to the Egg has the nymph and the sheet separating the yolk and the shell to which the nymph emerges from the connected oper and capitulum, sometimes when the nymph emerges, say the tropidoderus childrenii when the nymph hatches with the Egg still attached to the hind legs causing leg dis-formation.
Size	Size has been one of the most interesting aspects of cephalopod science to the general public Extinct taxa are also included.
Virginia opossum	The Virginia Opossum, commonly known as the North American Opossum, is the only marsupial found in North America north of the Rio Grande River. A solitary and nocturnal animal about the size of a domestic cat, it is a successful opportunist and is found throughout Central America and North America east of the Rockies from Costa Rica to southern Ontario (it was also introduced to California in 1910, and now occupies much of the Pacific coast); it seems to be still expanding its range northward. Its ancestors evolved in South America, but were enabled to invade North America in the Great American Interchange by the formation of the Isthmus of Panama about 3 million years ago.
Germ	The germ of a cereal is the reproductive part that germinates to grow into a plant; it is the embryo of the seed. Along with bran, germ is often a by-product of the milling that produces refined grain products. Cereal grains and their components, such as wheat germ, rice bran, and maize may be used as a source from which vegetable oil is extracted, or used directly as a food ingredient.
Heritability	In genetics, heritability is the proportion of phenotypic variation in a population that is attributable to genetic variation among individuals. Variation among individuals may be due to genetic and/or environmental factors. heritability analyses estimate the relative contributions of differences in genetic and non-genetic factors to the total phenotypic variance in a population.
Insects	Insects are arthropods, having a hard exoskeleton, a three-part body (head, thorax, and abdomen), three pairs of jointed legs, compound eyes, and two antennae. They are the most diverse group of animals on the planet and include approximately 2,200 species of praying mantis, 5,000 dragonfly, 20,000 grasshopper, 82,000 true bug, 120,000 fly, 110,000 bee, wasp, ant and sawfly, 170,000 butterfly and moth, and 360,000 beetle species described to date. The number of extant species is estimated at between six and ten million, with over a million species already described.
Mammal	Mammal s (formally mammal ia) are a class of vertebrate animals whose females are characterized by the possession of mammary glands while both males and females are characterized by sweat glands, hair, three middle ear bones used in hearing, and a neocortex region in the brain.

	mammal s are divided into three main categories depending how they are born. These categories are, monotremes, marsupials and placentals.
Species	In biology, a Species is: · a taxonomic rank (the basic rank of biological classification) or · a unit at that rank There are many definitions of what kind of unit a Species is (or should be.) A common definition is that of a group of organisms capable of interbreeding and producing fertile offspring, and separated from other such groups with which interbreeding does not (normally) happen. Other definitions may focus on similarity of DNA or morphology. Some Species are further subdivided into sub Species , and here also there is no close agreement on the criteria to be used.
Metabolism	Metabolism is the set of chemical reactions that occur in living organisms in order to maintain life. These processes allow organisms to grow and reproduce, maintain their structures, and respond to their environments. Metabolism is usually divided into two categories.
Fruit	The term fruit has different meanings dependent on context, and the term is not synonymous in food preparation and biology. fruit s are the means by which flowering plants disseminate seeds, and the presence of seeds indicates that a structure is most likely a fruit though not all seeds come from fruit s. No single terminology really fits the enormous variety that is found among plant fruit s.
Rat	Rat s are various medium-sized, long-tailed rodents of the superfamily Muroidea. "True Rat s" are members of the genus Rat tus, the most important of which to humans are the black Rat Rat tus Rat tus, and the brown Rat Rat tus norvegicus. Many members of other rodent genera and families are also called Rat s and share many characteristics with true Rat s.
Cell	The Cell is the structural and functional unit of all known living organisms. It is the smallest unit of an organism that is classified as living, and is often called the building block of life. Some organisms, such as most bacteria, are unicellular (consist of a single Cell.)
Cell division	Cell division is a process by which a cell, called the parent cell, divides into two or more cells, called daughter cells. cell division is usually a small segment of a larger cell cycle. This type of cell division in eukaryotes is known as mitosis, and leaves the daughter cell capable of dividing again.
Artificial selection	Artificial selection describes intentional breeding for certain traits, or combination of traits. It was defined by Charles Darwin in contrast to natural selection, in which the differential reproduction of organisms with certain traits is attributed to improved survival or reproductive ability ("Darwinian fitness") and the creation of new organisms, us humans breed them for specific traits. Artificial selection can also be unintentional; it is thought that domestication of crops by early humans was largely unintentional.

Protein

Protein s are organic compounds made of amino acids arranged in a linear chain. The amino acids in a polymer chain are joined together by the peptide bonds between the carboxyl and amino groups of adjacent amino acid residues. The sequence of amino acids in a protein is defined by the sequence of a gene, which is encoded in the genetic code.

Senescence

Senescence is a process induced by evolution into an organism"s genetic make up so that it may live to its healthiest until its reproductive age and die slowly and gradually thereafter. senescence encompasses all of the biological processes of a living organism"s approaching an advanced age (i.e., the combination of processes of deterioration which follow the period of development of an organism.) The word senescence is derived from the Latin word senex, meaning "old man" or "old age" or "advanced in age".

Telomere

A Telomere is a region of repetitive DNA at the end of chromosomes, which protects the end of the chromosome from destruction.

A Russian theorist Alexei Olovnikov was the first to recognize the problem of how chromosomes could replicate right to the tip, as such was impossible with replication in a 3" to 5" direction. To solve this and to accommodate Leonard Hayflick"s idea of limited somatic cell division, Olovnikov suggested that DNA sequences would be lost in every replicative phase until they reached a critical level, at which point cell division would stop.

Cancer

Cancer is a genetic disorder in which the normal control of cell growth is lost. Cancer genetics is now one of the fastest expanding medical specialties. At the molecular level, Cancer is caused by mutation(s) in DNA, which result in aberrant cell proliferation.

Nematode

The "roundworms" or " Nematode s" (phylum Nematoda) are the most diverse phylum of pseudocoelomates, and one of the most diverse of all animals. Nematode species are very difficult to distinguish; over 80,000 have been described, of which over 15,000 are parasitic. It has been estimated that the total number of described and undescribed roundworms might be more than 500,000.

Stem

A stem is one of two main structural axes of a vascular plant. The stem is normally divided into nodes and internodes, the nodes hold buds which grow into one or more leaves, inflorescence (flowers), cones or other stems etc. The internodes act as spaces that distance one node from another.

Stem cell

Stem cell s are cells found in most, if not all, multi-cellular organisms. They are characterized by the ability to renew themselves through mitotic cell division and differentiating into a diverse range of specialized cell types. Research in the Stem cell field grew out of findings by Canadian scientists Ernest A. McCulloch and James E. Till in the 1960s.

Tumor

A tumor or tumour is the name for a swelling or lesion formed by an abnormal growth of cells (termed neoplastic.) tumor is not synonymous with cancer. A tumor can be benign, pre-malignant or malignant, whereas cancer is by definition malignant.

Allele

An Allele) is one of a series of different forms of a gene. The word is a short from of allelomorph , which was used in the early days of genetics to describe variant forms of a gene detected as different phenotypes. Allele s are now understood to be alternative DNA sequences at the same physical gene locus, which may or may not result in different phenotypic traits.

Worm

The term Worm is used to describe many different distantly-related animals which have a long cylindrical body and no legs.

Most animals called " Worm s" are invertebrates, but the term is also used for the amphibian caecilians and the slow Worm Anguis, a legless burrowing lizard. Invertebrate animals commonly called " Worm s" include annelids (earth Worm s), nematodes (round Worm s), flat Worm s, marine polychaete Worm s (bristle Worm s), marine nemertean Worm ("bootlace Worm s"), and insect larvae such as caterpillars, grubs, and maggots.

Mutation

In biology, mutation s are changes to the nucleotide sequence of the genetic material of an organism. mutation s can be caused by copying errors in the genetic material during cell division, by exposure to ultraviolet or ionizing radiation, chemical mutagens, or viruses, or can be induced by the organism itself, by cellular processes such as hyper mutation . In multicellular organisms with dedicated reproductive cells, mutation s can be subdivided into germ line mutation s, which can be passed on to descendants through the reproductive cells, and somatic mutation s, which involve cells outside the dedicated reproductive group and which are not usually transmitted to descendants.

DNA

Deoxyribonucleic acid (DNA) is a nucleic acid that contains the genetic instructions used in the development and functioning of all known living organisms and some viruses. The main role of DNA molecules is the long-term storage of information. DNA is often compared to a set of blueprints or a recipe, or a code, since it contains the instructions needed to construct other components of cells, such as proteins and RNA molecules.

DNA mismatch repair

DNA mismatch repair is a system for recognizing and repairing erroneous insertion, deletion and mis-incorporation of bases that can arise during DNA replication and recombination, as well as repairing some forms of DNA damage .

Mismatch repair is strand-specific. During DNA synthesis the newly synthesised (daughter) strand will include errors commonly.

Gene

A Gene is the basic unit of heredity in a living organism. All living things depend on Gene s. Gene s hold the information to build and maintain their cells and pass Gene tic traits to offspring.

David

DAVID (the Database for Annotation, Visualization and Integrated Discovery) is a free online bioinformatics resources developed by the Laboratory of Immunopathogenesis and Bioinformatics (LIB.) All tools in the DAVID Bioinformatics Resources aim to provide functional interpretation of large lists of genes derived from genomic studies, e.g. microarray and proteomics studies. DAVID can be found at http://DAVID.niaid.nih.gov or http://DAVID.abcc.ncifcrf.gov

	The newly updated DAVID Bioinformatics Resources consists of the DAVID Knowledgebase and five integrated, web-based functional annotation tool suites: the DAVID Gene Functional Classification Tool, the DAVID Functional Annotation Tool, the DAVID Gene ID Conversion Tool, the DAVID Gene Name Viewer and the DAVID NIAID Pathogen Genome Browser.
Drosophila	Drosophila has long been a favorite model system for geneticists and developmental biologists studying embryogenesis. The small size, short generation time, and large brood size makes it ideal for genetic studies. Transparent embryos facilitate developmental studies.
Genetic	Genetics is the study of how living things receive common traits from previous generations. These traits are described by the Genetic information carried by a molecule called DNA. The instructions for constructing and operating an organism are contained in the organism"s DNA. Every living thing on earth has DNA in its cells. Genes are the hereditary components of DNA that occupy spots on chromosomes and determine characteristics in an organism.
Genetic drift	Genetic drift or allelic drift is the change in the relative frequency with which a gene variant (allele) occurs in a population that results from the fact that alleles in offspring are a random sample of those in the parents, and because of the role of chance in determining whether a given individual survives and reproduces. A population"s allele frequency is the fraction of the gene copies that share a particular form. Genetic drift is one of several evolutionary processes which lead to changes in allele frequencies over time.
Inbreeding depression	Inbreeding depression is reduced fitness in a given population as a result of breeding of related individuals. Breeding between closely related individuals, called inbreeding, results in more recessive deleterious traits manifesting themselves. The more closely related the breeding pair is, the more homozygous deleterious genes the offspring may have, resulting in very unfit individuals.
Indy	Indy, short for I"m not dead yet, is a gene of the model organism, the fruitfly Drosophila melanogaster. Mutant versions of this gene doubles the fruitflies" average lifespan (this is subject to controversy.) Its name originates from a repeated comic line in Monty Python and the Holy Grail .
Plants	Plants are living organisms belonging to the kingdom Plantae. They include familiar organisms such as trees, herbs, bushes, grasses, vines, ferns, mosses, and green algae. About 350,000 species of Plants, defined as seed Plants, bryophytes, ferns and fern allies, are estimated to exist currently.
Perennial Plant	A Perennial plant or perennial is a plant that lives for more than two years. When used by gardeners or horticulturalists, this term applies specifically to perennial herbaceous plants. Scientifically, woody plants like shrubs and trees are also perennial in their habit.

Predation	In ecology, Predation describes a biological interaction where a predator (an organism that is hunting) feeds on its prey, the organism that is attacked. Predators may or may not kill their prey prior to feeding on them, but the act of Predation always results in the death of the prey. The other main category of consumption is detritivory, the consumption of dead organic material (detritus.)
Acetylcholinesterase	Acetylcholinesterase is an enzyme that degrades the neurotransmitter acetylcholine, producing choline and an acetate group. It is mainly found at neuromuscular junctions and cholinergic synapses in the central nervous system, where its activity serves to terminate synaptic transmission. AChE has a very high catalytic activity -- each molecule of AChE degrades about 5000 molecules of acetylcholine per second.
Hunter-gatherer	A hunter-gatherer society is one whose primary subsistence method involves the direct procurement of edible plants and animals from the wild, foraging and hunting without significant recourse to the domestication of either. hunter-gatherer s obtain most from gathering rather than hunting; up to 80% of the food is obtained by gathering. The demarcation between hunter-gatherer s and other societies which rely more upon domestication is not a clear-cut one, as many contemporary societies use a combination of both strategies to obtain the foodstuffs required to sustain themselves.
Wasp	A Wasp is a predatory, flying, stinging insect, with a stinger and membranous forewings and hindwings. It is related to ants and bees, with all of them being members of order Hymenoptera, but is separated from ants and bees by having a stinger and no hair; bees have hair. A rough definition of the term Wasp is any member of the aculeate family Vespidae.
Altruism	Altruism is a well-documented animal behaviour, which appears most obviously in kin relationships but may also be evident amongst wider social groups. In the science of ethology (the study of behavior), and more generally in the study of social evolution, Altruism refers to behavior by an individual that increases the fitness of another individual while decreasing the fitness of the actor. Research in evolutionary theory has been applied to social behaviour, including Altruism.
Fish	FISH (fluorescence in situ hybridization) is a cytogenetic technique used to detect and localize the presence or absence of specific DNA sequences on chromosomes. FISH uses fluorescent probes that bind to only those parts of the chromosome with which they show a high degree of sequence similarity. Fluorescence microscopy can be used to find out where the fluorescent probe bound to the chromosomes.
Beetles	Beetles are the group of insects with the largest number of known species. They are classified in the order Coleoptera , which contains more described species than in any other order in the animal kingdom, constituting about 25% of all known life-forms. 40% of all described insect species are Beetles , and new species are frequently discovered.

Seed	A Seed, referred to as a kernel in some plants, is a small embryonic plant enclosed in a covering called the Seed coat, usually with some stored food. It is the product of the ripened ovule of gymnosperm and angiosperm plants which occurs after fertilization and some growth within the mother plant. The formation of the Seed completes the process of reproduction in Seed plants (started with the development of flowers and pollination), with the embryo developed from the zygote and the Seed coat from the integuments of the ovule.
Genomic	Genomic s is the study of the genomes of organisms. The field includes intensive efforts to determine the entire DNA sequence of organisms and fine-scale genetic mapping efforts. The field also includes studies of intra Genomic phenomena such as heterosis, epistasis, pleiotropy and other interactions between loci and alleles within the genome.
Genomic imprinting	Genomic imprinting is a genetic phenomenon by which certain genes are expressed in a parent-of-origin-specific manner. It is an inheritance process independent of the classical Mendelian inheritance. Imprinted genes are either expressed only from the allele inherited from the mother (eg.
Endosperm	Endosperm is the tissue produced in the seeds of most flowering plants around the time of fertilization. It surrounds the embryo and provides nutrition in the form of starch, though it can also contain oils and protein. This makes Endosperm an important source of nutrition in human diet.
Insulin-like growth factors	The Insulin-like growth factors are polypeptides with high sequence similarity to insulin. Insulin-like growth factorss are part of a complex system that cells use to communicate with their physiologic environment. This complex system (often referred to as the Insulin-like growth factors "axis") consists of two cell-surface receptors (Insulin-like growth factors1R and Insulin-like growth factors2R), two ligands (Insulin-like growth factors-1 and Insulin-like growth factors-2), a family of six high-affinity Insulin-like growth factors binding proteins (Insulin-like growth factorsBP 1-6), as well as associated Insulin-like growth factorsBP degrading enzymes, referred to collectively as proteases.
Coevolution	In a broad sense, biological coevolution is "the change of a biological object triggered by the change of a related object". coevolution can occur at multiple levels of biology: it can be as microscopic as correlated mutations between amino acids in a protein, or as macroscopic as covarying traits between different species in an environment. Each party in a coevolution ary relationship exerts selective pressures on the other, thereby affecting each others" evolution.
Assay	An Assay is a procedure in molecular biology for testing and/or measuring the activity of a drug or biochemical in an organism or organic sample. A quantitative Assay may also measure the amount of a substance in a sample. Bio Assay s and immuno Assay s are among the many varieties of specialized biochemical Assay s.
Genetic variation	Genetic variation, variation in alleles of genes, occurs both within and among populations. genetic variation is important because it provides the "raw material" for natural selection.

genetic variation among individuals within a population can be identified at a variety of levels.

Maintenance

Maintenance of an organism is the collection of processes to stay alive, excluding production processes. The Dynamic Energy Budget theory delineates two classes

· Somatic Maintenance. This comprises the turnover of structural mass (mainly proteins), the Maintenance of concentration gradients of metabolites across membranes, activity
· Maturity Maintenance. This comprises the Maintenance of defence systems (such as the immune system), the preparation of the body for reproduction.

The theory assumes that maturity Maintenance costs can be reduced more easily during starvation than somatic Maintenance costs. Under extreme starvation conditions, somatic Maintenance costs are paid from structural mass, which causes shrinking.Some organism manage to switch to the turpor state under starvation conditions, and reduce their Maintenance costs.

Trait

A trait is a distinct variant of a phenotypic character of an organism that may be inherited, environmentally determined or somewhere in between. For example, eye color is a character or abstraction of an attribute, while blue, brown and hazel are traits.
A trait may be any single feature or quantifiable measurement of an organism.

Sea urchins

Sea urchins or urchins are small, spiny, globular animals that compose part of class Echinoidea. They are found in oceans all over the world. Their shell, or "test", is round and spiny, typically from 3 to 10 cm across.

Form

In botanical nomenclature, a Form is a low-level taxonomic rank below that of variety; it is an infraspecific taxon. Its name consists of three parts: a genus name, a specific epithet, and an infraspecific epithet. The abbreviation "f." or the full "forma" should be put before the infraspecific epithet to indicate the rank.

Larva

A Larva is a young form of animal with indirect development, going through or undergoing metamorphosis (for example, insects, amphibians, or cnidarians.)
The Larva can look completely different from the adult form, for example, a caterpillar differs from a butterfly. Larva e often have special (Larva l) organs which do not occur in the adult form.

Group

In naming cultivated plants, a Group is a formal classification category, under the International Code of Nomenclature for Cultivated Plants (ICNCP):

> ICNCP Art. 3.1: "a formal category for assembing cultivars, individual plants or assemblages of plants on the basis of defined similarity" .

The term "Group" (with a capital G) was introduced in the 2004 ICNCP, replacing the "Cultivar-Group" of the 1995 ICNCP.

A Group is united by some common trait; for example there may be a Group of yellow-flowering cultivars, a Group of cultivars with variegated leaves, a Group of cultivars resistant to a particular disease, etc.

Gill	A Gill is an anatomical structure found in many aquatic organisms. It is a respiration organ whose function is the extraction of oxygen from water and the excretion of carbon dioxide. The microscopic structure of a Gill is such that it presents a very large surface area to the external environment.
Balanus glandula	Balanus glandula, commonly called white acorn barnacle, is one of the most common barnacle species on the Pacific coast of North America, distributed from the U.S. state of Alaska to the border between California and the Mexican state of Baja California, commonly found in intertidal waters on mussels, rocks and pier pilings. It is a moderate-sized barnacle with a length of about 15-20 mm. The shell is formed by overlapping plates.
Barnacle	A Barnacle is a type of arthropod belonging to infraclass Cirripedia in the subphylum Crustacea, and is hence related to crabs and lobsters. Barnacle s are exclusively marine, and tend to live in shallow and tidal waters, typically in erosive settings. They are sessile suspension feeders, and have two nektonic larval stages.

Antibiotic resistance	Antibiotic resistance is the ability of a microorganism to withstand the effects of antibiotics. It is a specific type of drug resistance. antibiotic resistance evolves via natural selection acting upon random mutation, but it can also be engineered by applying an evolutionary stress on a population.
Cholera	Cholera sometimes known as Asiatic or epidemic Cholera is an infectious gastroenteritis caused by enterotoxin-producing strains of the bacterium Vibrio Cholera e. Transmission to humans occurs through eating food or drinking water contaminated with Vibrio Cholera e from other Cholera patients. The major reservoir for Cholera was long assumed to be humans themselves, but considerable evidence exists that aquatic environments can serve as reservoirs of the bacteria.
Horse murders	The Horse murders scandal was a form of insurance fraud in the United States in which expensive horses, many of them show jumpers, were insured against death, accident and then killed to collect the insurance money. It is not known how many horses were murdered between the mid 1970s and the mid-1990s, when a Federal Bureau of Investigation (FBI) investigation brought the horse killings to light, but the number is thought to be well over 50, and may have been as high as 100. In addition, in 1977, the heiress Helen Brach disappeared and was presumed by law enforcement agents to have been murdered by the perpetrators of these crimes, because she threatened to report their criminal activity to authorities; continuing investigations into Brach"s death began to uncover the insurance fraud in the 1990s.
Germ	The germ of a cereal is the reproductive part that germinates to grow into a plant; it is the embryo of the seed. Along with bran, germ is often a by-product of the milling that produces refined grain products. Cereal grains and their components, such as wheat germ, rice bran, and maize may be used as a source from which vegetable oil is extracted, or used directly as a food ingredient.
Macrophages	Macrophages are white blood cells within tissues, produced by the division of monocytes. Human Macrophages are about 21 micrometres in diameter. Monocytes and Macrophages are phagocytes, acting in both non-specific defense as well as to help initiate specific defense mechanisms (or adaptive immunity) of vertebrate animals.
Vibrio	Vibrio is a genus of Gram-negative bacteria possessing a curved rod shape. Typically found in saltwater, Vibrio are facultative anaerobes that test positive for oxidase and do not form spores. All members of the genus are motile and have polar flagella with sheaths.
Vibrio cholerae	Vibrio cholerae is a motile gram negative curved-rod shaped bacterium with a polar flagellum that causes cholera in humans. V. cholerae and other species of the genus Vibrio belong to the gamma subdivision of the Proteobacteria. There are two major strains of V. cholerae, classic and El Tor, and numerous other serogroups.

Bacteria

The Bacteria are a large group of unicellular microorganisms. Typically a few micrometres in length, Bacteria have a wide range of shapes, ranging from spheres to rods and spirals. Bacteria are ubiquitous in every habitat on Earth, growing in soil, acidic hot springs, radioactive waste, water, and deep in the Earth"s crust, as well as in organic matter and the live bodies of plants and animals.

Charles Robert Darwin

Charles Robert Darwin FRS (12 February 1809 - 19 April 1882) was an English naturalist who realised and presented compelling evidence that all species of life have evolved over time from common ancestors, through the process he called natural selection. The fact that evolution occurs became accepted by the scientific community and much of the general public in his lifetime, while his theory of natural selection came to be widely seen as the primary explanation of the process of evolution in the 1930s, and now forms the basis of modern evolutionary theory. In modified form, Darwin"s scientific discovery is the unifying theory of the life sciences, providing logical explanation for the diversity of life.

Species

In biology, a Species is:

· a taxonomic rank (the basic rank of biological classification) or
· a unit at that rank

There are many definitions of what kind of unit a Species is (or should be.) A common definition is that of a group of organisms capable of interbreeding and producing fertile offspring, and separated from other such groups with which interbreeding does not (normally) happen. Other definitions may focus on similarity of DNA or morphology. Some Species are further subdivided into sub Species , and here also there is no close agreement on the criteria to be used.

Tuberculosis

Tuberculosis is a common and often deadly infectious disease caused by mycobacteria, in humans mainly Mycobacterium Tuberculosis . Tuberculosis usually attacks the lungs (as pulmonary TB) but can also affect the central nervous system, the lymphatic system, the circulatory system, the genitourinary system, the gastrointestinal system, bones, joints, and even the skin. Other mycobacteria such as Mycobacterium bovis, Mycobacterium africanum, Mycobacterium canetti, and Mycobacterium microti also cause Tuberculosis, but these species are less common in humans.

Pathogen

A pathogen "I give birth to"), infectious agent is a biological agent that causes disease or illness to its host. There are several substrates and pathways whereby pathogen s can invade a host; the principal pathways have different episodic time frames, but soil contamination has the longest or most persistent potential for harboring a pathogen

The body contains many natural defenses against some of the common pathogen s (such as Pneumocystis) in the form of the human immune system and by some "helpful" bacteria present in the human body"s normal flora.

Nucleotide	Nucleotide s are molecules that, when joined together, make up the structural units of RNA and DNA. Additionally, Nucleotide s play central roles in metabolism. In that capacity, they serve as sources of chemical energy (adenosine triphosphate and guanosine triphosphate), participate in cellular signaling (cyclic guanosine monophosphate and cyclic adenosine monophosphate), and are incorporated into important cofactors of enzymatic reactions (coenzyme A, flavin adenine di Nucleotide , flavin mono Nucleotide , and nicotinamide adenine di Nucleotide phosphate.) Figure 1: Structural elements of the most common Nucleotide s Figure 2: Ribose structure indicating numbering of carbon atoms A Nucleotide is composed of a nucleobase (nitrogenous base) and a five-carbon sugar (either ribose or 2"-deoxyribose), and one to three phosphate groups.
Sialic acid	Sialic acid is a generic term for the N- or O-substituted derivatives of neuraminic acid, a monosaccharide with a nine-carbon backbone. It is also the name for the most common member of this group, N-acetylneuraminic acid (Neu5Ac or NANA.) Sialic acid s are found widely distributed in animal tissues and to a lesser extent in other species ranging from plants and fungi to yeasts and bacteria, mostly in glycoproteins and gangliosides.
Leukemia inhibitory factor	Leukemia inhibitory factor an interleukin 6 class cytokine, is a chemical in cells that affects their growth and development. Leukemia inhibitory factor derives its name from its ability to induce the terminal differentiation of myeloid leukaemic cells. Other properties attributed to the cytokine include: the growth promotion and cell differentiation of different types of target cells, influence on bone metabolism, cachexia, neural development, embryogenesis and inflammation.
DNA	Deoxyribonucleic acid (DNA) is a nucleic acid that contains the genetic instructions used in the development and functioning of all known living organisms and some viruses. The main role of DNA molecules is the long-term storage of information. DNA is often compared to a set of blueprints or a recipe, or a code, since it contains the instructions needed to construct other components of cells, such as proteins and RNA molecules.
DNA sequence	A DNA sequence or genetic sequence is a succession of letters representing the primary structure of a real or hypothetical DNA molecule or strand, with the capacity to carry information as described by the central dogma of molecular biology. The possible letters are A, C, G, and T, representing the four nucleotide bases of a DNA strand -- adenine, cytosine, guanine, thymine -- covalently linked to a phosphodiester backbone. In the typical case, the sequences are printed abutting one another without gaps, as in the sequence AAAGTCTGAC, read left to right in the 5" to 3" direction.

Neutral theory of molecular evolution	The Neutral theory of molecular evolution is an influential theory, which was introduced with effect by Motoo Kimura in the late 1960s and early 1970s. The theory states that the vast majority of evolutionary changes at the molecular level are caused by random drift of selectively neutral mutants. Although the theory was received by some as an argument against Darwin"s theory of evolution by natural selection, Kimura maintained, and most modern evolutionary biologists agree, that the two theories are compatible: "The theory does not deny the role of natural selection in determining the course of adaptive evolution" (Kimura, 1986.)
Virus	A virus is a microscopic infectious agent that can reproduce only inside a host cell. virus es infect all types of organisms: from animals and plants, to bacteria and archaea. Since the initial discovery of tobacco mosaic virus by Martinus Beijerinck in 1898, more than 5,000 types of virus have been described in detail, although most types of virus remain undiscovered.
Vaccine	A vaccine is a biological preparation that improves immunity to a particular disease. A vaccine typically contains a small amount of an agent that resembles a microorganism. The agent stimulates the body"s immune system to recognize the agent as foreign, destroy it, and "remember" it, so that the immune system can more easily recognize and destroy any of these microorganisms that it later encounters.
Neuraminidase	Neuraminidase enzymes are glycoside hydrolase enzymes (EC 3.2.1.18) which cleave the glycosidic linkages of neuraminic acid. Neuraminidase enzymes are a large family, found in a range of organisms. The most commonly known Neuraminidase is the viral Neuraminidase, a drug target for the prevention of influenza infection.
Mycobacterium	Mycobacterium is a genus of Actinobacteria, given its own family, the Mycobacteriaceae. The genus includes pathogens known to cause serious diseases in mammals, including tuberculosis and leprosy. The Latin prefix "myco--" means both fungus and wax; its use here relates to the "waxy" compounds in the cell wall.
Gene	A Gene is the basic unit of heredity in a living organism. All living things depend on Gene s. Gene s hold the information to build and maintain their cells and pass Gene tic traits to offspring.
Plasmid	A Plasmid is an extra-chromosomal DNA molecule separate from the chromosomal DNA which is capable of replicating independently of the chromosomal DNA. In many cases, it is circular and double-stranded. Plasmid s usually occur naturally in bacteria, but are sometimes found in eukaryotic organisms (e.g., the 2-micrometre-ring in Saccharomyces cerevisiae.)
Bacteriophage	A Bacteriophage is any one of a number of viruses that infect bacteria. Bacteriophage s are among the most common organisms on Earth. The term is commonly used in its shortened form, phage.

Human	A Human is a member of a species of bipedal primates in the family Hominidae . DNA and fossil evidence indicates that modern Human s originated in east Africa about 200,000 years ago. When compared to other animals and primates, Human s have a highly developed brain, capable of abstract reasoning, language, introspection and problem solving.
Adenosine	Adenosine is a nucleoside composed of a molecule of adenine attached to a ribose sugar molecule (ribofuranose) moiety via a β-N_9-glycosidic bond. Adenosine plays an important role in biochemical processes, such as energy transfer--as Adenosine triphosphate (ATP) and Adenosine diphosphate (ADP)--as well as in signal transduction as cyclic Adenosine monophosphate, cAMP. It is also an inhibitory neurotransmitter, believed to play a role in promoting sleep and suppressing arousal, with levels increasing with each hour an organism is awake. Adenosine is an endogenous purine nucleoside that modulates many physiological processes.
Adenosine deaminase deficiency	Adenosine deaminase deficiency is an autosomal recessive metabolic disorder that causes immunodeficiency. It accounts for about 15% of all cases of severe combined immunodeficiency ADA deficiency may present in infancy, childhood, adolescense, or adulthood.
Back	The Back describes the area of horse anatomy where the saddle goes, and in popular usage extends to include the loin or lumbar region behind the thoracic vertebrae that also is crucial to a horse"s weight-carrying ability. These two sections of the vertebral column beginning at the withers, the start of the thoracic vertebrae, and extend to the last lumbar vertebra. Because horses are ridden by humans, the strength and structure of the horse"s Back is critical to the animal"s usefulness.
Mutation	In biology, mutation s are changes to the nucleotide sequence of the genetic material of an organism. mutation s can be caused by copying errors in the genetic material during cell division, by exposure to ultraviolet or ionizing radiation, chemical mutagens, or viruses, or can be induced by the organism itself, by cellular processes such as hyper mutation . In multicellular organisms with dedicated reproductive cells, mutation s can be subdivided into germ line mutation s, which can be passed on to descendants through the reproductive cells, and somatic mutation s, which involve cells outside the dedicated reproductive group and which are not usually transmitted to descendants.
Cell	The Cell is the structural and functional unit of all known living organisms. It is the smallest unit of an organism that is classified as living, and is often called the building block of life. Some organisms, such as most bacteria, are unicellular (consist of a single Cell.)

Population	In biology, a population is the collection of inter-breeding organisms of a particular species; in sociology, a collection of human beings. Individuals within a population share a factor may be reduced by statistical means, but such a generalization may be too vague to imply anything. Demography is used extensively in marketing, which relates to economic units, such as retailers, to potential customers.
Enzymes	Enzymes are biomolecules that catalyze (i.e., increase the rates of) chemical reactions. Nearly all known Enzymes are proteins. However, certain RNA molecules can be effective biocatalysts too.
Cancer	Cancer is a genetic disorder in which the normal control of cell growth is lost. Cancer genetics is now one of the fastest expanding medical specialties. At the molecular level, Cancer is caused by mutation(s) in DNA, which result in aberrant cell proliferation.
Gene therapy	Gene therapy is the insertion of genes into an individual"s cells and tissues to treat a disease, such as a hereditary disease in which a deleterious mutant allele is replaced with a functional one. Although the technology is still in its infancy, it has been used with some success. Antisense therapy is not strictly a form of Gene therapy, but is a genetically-mediated therapy and is often considered together with other methods.
Genetic	Genetics is the study of how living things receive common traits from previous generations. These traits are described by the Genetic information carried by a molecule called DNA. The instructions for constructing and operating an organism are contained in the organism"s DNA. Every living thing on earth has DNA in its cells. Genes are the hereditary components of DNA that occupy spots on chromosomes and determine characteristics in an organism.
Genetic variation	Genetic variation, variation in alleles of genes, occurs both within and among populations. genetic variation is important because it provides the "raw material" for natural selection. genetic variation among individuals within a population can be identified at a variety of levels.
Tumor	A tumor or tumour is the name for a swelling or lesion formed by an abnormal growth of cells (termed neoplastic.) tumor is not synonymous with cancer. A tumor can be benign, pre-malignant or malignant, whereas cancer is by definition malignant.
Hunter-gatherer	A hunter-gatherer society is one whose primary subsistence method involves the direct procurement of edible plants and animals from the wild, foraging and hunting without significant recourse to the domestication of either. hunter-gatherer s obtain most from gathering rather than hunting; up to 80% of the food is obtained by gathering. The demarcation between hunter-gatherer s and other societies which rely more upon domestication is not a clear-cut one, as many contemporary societies use a combination of both strategies to obtain the foodstuffs required to sustain themselves.
Adaptation	In ocular physiology, Adaptation is the ability of the eye to adjust to various levels of darkness and light.

	The human eye can function from very dark to very bright levels of light -- its sensing capabilities reach across nine orders of magnitude. This means that the brightest and the darkest light signal that the eye can sense are a factor of roughly one thousand million apart.
Altruism	Altruism is a well-documented animal behaviour, which appears most obviously in kin relationships but may also be evident amongst wider social groups. In the science of ethology (the study of behavior), and more generally in the study of social evolution, Altruism refers to behavior by an individual that increases the fitness of another individual while decreasing the fitness of the actor. Research in evolutionary theory has been applied to social behaviour, including Altruism.
Aeromonas	Aeromonas is a gram-negative, facultative anaerobic rod that morphologically resembles members of the family Enterobacteriaceae. Fourteen species of Aeromonas have been described, most of which have been associated with human diseases. The most important pathogens are A. hydrophila, A. caviae, and A. veronii biovar sobria.
Aeromonas hydrophila	Aeromonas hydrophila is a heterotrophic, gram-negative bacterium, mainly found in areas with a warm climate. This bacterium can also be found in fresh, salt, marine, estuarine, chlorinated, and un-chlorinated water. Aeromonas hydrophila can survive in aerobic and anaerobic environments.
Thermoregulation	Thermoregulation is the ability of an organism to keep its body temperature within certain boundaries, even when the surrounding temperature is very different. This process is one aspect of homeostasis: a dynamic state of stability between an animal"s internal environment and its external environment (the study of such processes in zoology has been called ecophysiology or physiological ecology.) If the body is unable to maintain a normal temperature and it increases significantly above normal, a condition known as heat stroke occurs.
Salmonella	Salmonella is a genus of rod-shaped, Gram-negative, non-spore forming, predominantly motile enterobacteria with diameters around 0.7 to 1.5 Âµm, lengths from 2 to 5 Âµm, and flagella which project in all directions (i.e. peritrichous.) They are chemoorganotrophs, obtaining their energy from oxidation and reduction reactions using organic sources and are facultative anaerobes; most species produce hydrogen sulfide, which can readily be detected by growing them on media containing ferrous sulfate, such as TSI. Most isolates exist in two phases; phase I is the motile phase and phase II the non-motile phase. Cultures that are non-motile upon primary culture may be swithched to the motile phase using a Craigie tube.
Cold	ColD is a PLP-dependent enzyme responsible for the removal of the C-3" hydroxyl group during the biosynthesis of GDP-colitose.

Genetic determinism	Genetic determinism is the belief that genes determine physical and behavioral phenotypes. It is usually taken to mean "that the genotype completely determines the phenotype, that is, the genes completely determine how an organism turns out", or that "genes alone determine human traits and behaviours." The term may be applied to the mapping of a single gene to a single phenotype, or more widely to the discredited belief that most or all phenotypes are determined exclusively by genes. While it is well established that most phenotypic variability is strongly influenced by genes, it is clear that environment plays an important role, and non-genetic mechanisms of inheritance are also known to exist.
Acetylcholinesterase	Acetylcholinesterase is an enzyme that degrades the neurotransmitter acetylcholine, producing choline and an acetate group. It is mainly found at neuromuscular junctions and cholinergic synapses in the central nervous system, where its activity serves to terminate synaptic transmission. AChE has a very high catalytic activity -- each molecule of AChE degrades about 5000 molecules of acetylcholine per second.

C-value	The term C-value refers to the amount of DNA contained within a haploid nucleus (e.g., in a gamete or one half the amount in a diploid somatic cell) of a eukaryotic organism. In some cases (notably among diploid organisms), the terms C-value and genome size are used interchangeably, however in polyploids the C-value may represent two genomes contained within the same nucleus. Greilhuber et al.
DNA	Deoxyribonucleic acid (DNA) is a nucleic acid that contains the genetic instructions used in the development and functioning of all known living organisms and some viruses. The main role of DNA molecules is the long-term storage of information. DNA is often compared to a set of blueprints or a recipe, or a code, since it contains the instructions needed to construct other components of cells, such as proteins and RNA molecules.
Genome	In classical genetics, the Genome of a diploid organism including eukarya refers to a full set of chromosomes or genes in a gamete; thereby, a regular somatic cell contains two full sets of Genome s. In haploid organisms, including bacteria, archaea, viruses, and mitochondria, a cell contains only a single set of the Genome usually in a single circular or contiguous linear DNA (or RNA for retroviruses.) In modern molecular biology the Genome of an organism is its hereditary information encoded in DNA (or, for retroviruses, RNA.)
Plasmid	A Plasmid is an extra-chromosomal DNA molecule separate from the chromosomal DNA which is capable of replicating independently of the chromosomal DNA. In many cases, it is circular and double-stranded. Plasmid s usually occur naturally in bacteria, but are sometimes found in eukaryotic organisms (e.g., the 2-micrometre-ring in Saccharomyces cerevisiae.)
Genomic	Genomic s is the study of the genomes of organisms. The field includes intensive efforts to determine the entire DNA sequence of organisms and fine-scale genetic mapping efforts. The field also includes studies of intra Genomic phenomena such as heterosis, epistasis, pleiotropy and other interactions between loci and alleles within the genome.
Mutation	In biology, mutation s are changes to the nucleotide sequence of the genetic material of an organism. mutation s can be caused by copying errors in the genetic material during cell division, by exposure to ultraviolet or ionizing radiation, chemical mutagens, or viruses, or can be induced by the organism itself, by cellular processes such as hyper mutation . In multicellular organisms with dedicated reproductive cells, mutation s can be subdivided into germ line mutation s, which can be passed on to descendants through the reproductive cells, and somatic mutation s, which involve cells outside the dedicated reproductive group and which are not usually transmitted to descendants.
Alu sequence	An Alu sequence is a short stretch of DNA originally characterized by the action of the Alu restriction endonuclease. Alu sequence s of different kinds occur in large numbers in primate genomes. In fact, Alu sequence s are the most abundant mobile elements in the human genome.

Human	A Human is a member of a species of bipedal primates in the family Hominidae . DNA and fossil evidence indicates that modern Human s originated in east Africa about 200,000 years ago. When compared to other animals and primates, Human s have a highly developed brain, capable of abstract reasoning, language, introspection and problem solving.
Hollywood	A Hollywood is a drug slang term referring to an unusually large and often dangerous single dosage of cocaine. Depending on a user"s experience with and tolerance of the drug, a typical line measures between 50-75mg of cocaine. However, a Hollywood line can exceed 200 mg of cocaine in a single dosage.
Long terminal repeats	Long terminal repeats are seqences of DNA that repeat hundreds or thousands of times. They are found in retroviral DNA, flanking functional genes. They are used by viruses to insert their genetic sequences into the host genomes.
Retrotransposons	Retrotransposons are genetic elements that can amplify themselves in a genome and are ubiquitous components of the DNA of many eukaryotic organisms. They are a subclass of transposon. They are particularly abundant in plants, where they are often a principal component of nuclear DNA. In maize, 49-78% of the genome is made up of Retrotransposons.
Saccharomyces	Saccharomyces is a genus in the kingdom of fungi that includes many species of yeast. Saccharomyces is from Latin meaning sugar fungi. Many members of this genus are considered very important in food production.
Saccharomyces cerevisiae	Saccharomyces cerevisiae is a species of budding yeast. It is perhaps the most useful yeast owing to its use since ancient times in baking and brewing. It is believed that it was originally isolated from the skins of grapes
Acid	An acid is traditionally considered any chemical compound that, when dissolved in water, gives a solution with a hydrogen ion activity greater than in pure water, i.e. a pH less than 7.0. That approximates the modern definition of Johannes Nicolaus Brønsted and Martin Lowry, who independently defined an acid as a compound which donates a hydrogen ion (H^+) to another compound (called a base.) Common examples include acetic acid and sulfuric acid (used in car batteries.)
Virus	A virus is a microscopic infectious agent that can reproduce only inside a host cell. virus es infect all types of organisms: from animals and plants, to bacteria and archaea. Since the initial discovery of tobacco mosaic virus by Martinus Beijerinck in 1898, more than 5,000 types of virus have been described in detail, although most types of virus remain undiscovered.

Insertion sequence	An Insertion sequence is a short DNA sequence that acts as a simple transposable element. Insertion sequence s have two major characteristics: they are small relative to other transposable elements and only code for proteins implicated in the transposition activity These proteins are usually the transposase which catalyses the enzymatic reaction allowing the Insertion sequence to move, and also one regulatory protein which either stimulates or inhibits the transposition activity.
P element	A P element is a transposon that is present in the fruit fly Drosophila melanogaster and is used widely for mutagenesis and the creation of genetically modified flies used for genetic research. The P element gives rise to a phenotype known as hybrid dysgenesis P element s seem to have first appeared in the species only in the middle of the twentieth century.
Transposase	Transposase is an enzyme that binds to the ends of a transposon and catalyzes the movement of the transposon to another part of the genome by a cut and paste mechanism or a replicative transposition mechanism. The word "Transposase" was first coined by the individuals who cloned the enzyme required for tranposition of the Tn3 transposon . The existence of transposons was postulated in the late 1940s by Barbara McClintock, who was studying the inheritance of maize, however the actual molecular basis for transposition was described by later groups.
Transposons	Transposons are sequences of DNA that can move around to different positions within the genome of a single cell, a process called transposition. In the process, they can cause mutations and change the amount of DNA in the genome. Transposons were also once called "jumping genes", and are examples of mobile genetic elements.
Genetic	Genetics is the study of how living things receive common traits from previous generations. These traits are described by the Genetic information carried by a molecule called DNA. The instructions for constructing and operating an organism are contained in the organism"s DNA. Every living thing on earth has DNA in its cells. Genes are the hereditary components of DNA that occupy spots on chromosomes and determine characteristics in an organism.
Gene	A Gene is the basic unit of heredity in a living organism. All living things depend on Gene s. Gene s hold the information to build and maintain their cells and pass Gene tic traits to offspring.
Genetic recombination	Genetic recombination is the process by which a strand of genetic material (usually DNA; but can also be RNA) is broken and then joined to a different DNA molecule. In eukaryotes recombination commonly occurs during meiosis as chromosomal crossover between paired chromosomes. This process leads to offspring having different combinations of genes from their parents and can produce new chimeric alleles.

Methylation	Methylation in the chemical sciences denotes the attachment or substitution of a methyl group on various substrates. This term is commonly used in chemistry, biochemistry, soil science and the biological sciences. In biochemistry, Methylation more specifically refers to the replacement of a hydrogen atom with the methyl group.
Selfish gene	The Selfish Gene is a book on evolution by Richard Dawkins, published in 1976. It builds upon the principal theory of George C. Williams"s first book Adaptation and Natural Selection. Dawkins coined the term "Selfish Gene" as a way of expressing the gene-centred view of evolution, which holds that evolution is best viewed as acting on genes and that selection at the level of organisms or populations almost never overrides selection based on genes.
Chromosome	A Chromosome is an organized structure of DNA and protein that is found in cells. It is a single piece of coiled DNA containing many genes, regulatory elements and other nucleotide sequences. Chromosome s also contain DNA-bound proteins, which serve to package the DNA and control its functions.
Inversion	An inversion is a chromosome rearrangement in which a segment of a chromosome is reversed end to end. An inversion occurs when a single chromosome undergoes breakage and rearrangement within itself. Inversions are of two types: paracentric and pericentric.
DNA methylation	DNA methylation is a type of chemical modification of DNA that can be inherited and subsequently removed without changing the original DNA sequence. As such, it is part of the epigenetic code and is also the best characterized epigenetic mechanism. Because methylation is a common capability of all viruses for self non-self identification, the epigenetic code could be a persistent remnant of ancient viral infection events.
Protein	Protein s are organic compounds made of amino acids arranged in a linear chain. The amino acids in a polymer chain are joined together by the peptide bonds between the carboxyl and amino groups of adjacent amino acid residues. The sequence of amino acids in a protein is defined by the sequence of a gene, which is encoded in the genetic code.
Transposon Mutagenesis	Transposon Mutagenesis mutagenesis is a biology process that allows genes to be transferred to a host organism"s chromosome, interrupting or modifying the function of an extant gene on the chromosome and causing mutation. Transposon Mutagenesis was first studied by Barbara McClintock in the mid-20th century during her Nobel Prize-winning work with corn. In the case of bacteria, transposition mutagenesis is usually accomplished by way of a plasmid from which a transposon is extracted and inserted into the host chromosome.

Telomerase	Telomerase is an enzyme that adds specific DNA sequence repeats ("TTAGGG" in all vertebrates) to the 3" end of DNA strands in the telomere regions, which are found at the ends of eukaryotic chromosomes. The telomeres contain condensed DNA material, giving stability to the chromosomes. The enzyme is a reverse transcriptase that carries its own RNA molecule, which is used as a template when it elongates telomeres, which are shortened after each replication cycle.
Archaeoglobus	In taxonomy, Archaeoglobus is a genus of the Archaeoglobaceae. The genus Archaeoglobus is a hyperthermophilic member of the Euryarchaeota. It is composed of two species, A. fulgidus and A. profundus which were isolated from hydrothermal vents.
Antibiotic resistance	Antibiotic resistance is the ability of a microorganism to withstand the effects of antibiotics. It is a specific type of drug resistance. antibiotic resistance evolves via natural selection acting upon random mutation, but it can also be engineered by applying an evolutionary stress on a population.
Chloroplast	Chloroplast s are organelles found in plant cells and other eukaryotic organisms that conduct photosynthesis. Chloroplast s capture light energy to conserve free energy in the form of ATP and reduce NADP to NADPH through a complex set of processes called photosynthesis. The word Chloroplast is derived from the Greek words chloros which means green and plast which means form or entity.
Influenza	Influenza, commonly referred to as the flu, is an infectious disease caused by RNA viruses of the family Orthomyxoviridae (the Influenza viruses), that affects birds and mammals. The name Influenza comes from the Italian Influenza, meaning "influence" . The most common symptoms of the disease are chills, fever, sore throat, muscle pains, severe headache, coughing, weakness and general discomfort.
Transduction	Transduction is the process by which DNA is transferred from one bacterium to another by a virus. It also refers to the process whereby foreign DNA is introduced into another cell via a viral vector. This is a common tool used by molecular biologists to stably introduce a foreign gene into a host cell"s genome.
Archaea	The Archaea are a group of single-celled microorganisms. A single individual or species from this domain is called an archaeon (sometimes spelled "archeon".) They have no cell nucleus or any other organelles within their cells.
Bacteria	The Bacteria are a large group of unicellular microorganisms. Typically a few micrometres in length, Bacteria have a wide range of shapes, ranging from spheres to rods and spirals. Bacteria are ubiquitous in every habitat on Earth, growing in soil, acidic hot springs, radioactive waste, water, and deep in the Earth"s crust, as well as in organic matter and the live bodies of plants and animals.

Horse murders

The Horse murders scandal was a form of insurance fraud in the United States in which expensive horses, many of them show jumpers, were insured against death, accident and then killed to collect the insurance money. It is not known how many horses were murdered between the mid 1970s and the mid-1990s, when a Federal Bureau of Investigation (FBI) investigation brought the horse killings to light, but the number is thought to be well over 50, and may have been as high as 100. In addition, in 1977, the heiress Helen Brach disappeared and was presumed by law enforcement agents to have been murdered by the perpetrators of these crimes, because she threatened to report their criminal activity to authorities; continuing investigations into Brach''s death began to uncover the insurance fraud in the 1990s.

Cyanobacteria

Cyanobacteria blue-green bacteria or Cyanophyta, is a phylum of bacteria that obtain their energy through photosynthesis. The name "Cyanobacteria" comes from the color of the bacteria . They are a significant component of the marine nitrogen cycle and an important primary producer in many areas of the ocean, but are also found in habitats other than the marine environment; in particular Cyanobacteria are known to occur in both freshwater, hypersaline inland lakes and in arid areas where they are a major component of biological soil crusts.

Mitochondrial DNA

Mitochondrial DNA is the DNA located in organelles called mitochondria. Most other DNA present in eukaryotic organisms is found in the cell nucleus. Mitochondrial DNA was discovered by Margit M. K. Nass and Sylvan Nass by electron microscopy as DNAase-sensitive thread inside mitochondria, and by Ellen Haslbrunner, Hans Tuppy and Gottfried Schatz by biochemical assays on highly purified mitochondrial fractions.

Organelle

In cell biology, an Organelle is a specialized subunit within a cell that has a specific function, and is usually separately enclosed within its own lipid membrane. A typical animal cell. Within the cytoplasm, the major Organelle s and cellular structures include: (1) nucleolus (2) nucleus (3) ribosome (4) vesicle (5) rough endoplasmic reticulum (6) Golgi apparatus (7) cytoskeleton (8) smooth endoplasmic reticulum (9) mitochondria (10) vacuole (11) cytosol (12) lysosome (13) centriole.

The name Organelle comes from the idea that these structures are to cells what an organ is to the body (hence the name Organelle the suffix -elle being a diminutive.)

Proteobacteria

The Proteobacteria are a major group (phylum) of bacteria.

They include a wide variety of pathogens, such as Escherichia, Salmonella, Vibrio, Helicobacter, and many other notable genera. Others are free-living, and include many of the bacteria responsible for nitrogen fixation.

Ribulose

Ribulose is a ketopentose -- a monosaccharide containing five carbon atoms, and including a ketone functional group. It has chemical formula $C_5H_{10}O_5$. Two enantiomers are possible, D-Ribulose and L-Ribulose (L-erythro-pentulose.)

Messenger ribonucleic acid	Messenger ribonucleic acid is a molecule of RNA encoding a chemical "blueprint" for a protein product. mRNA is transcribed from a DNA template, and carries coding information to the sites of protein synthesis: the ribosomes. Here, the nucleic acid polymer is translated into a polymer of amino acids: a protein.
Locus	In the fields of genetics and evolutionary computation, a Locus is a fixed position on a chromosome such as the position of a biomarker that may be occupied by one or more genes. A variant of the DNA sequence at a given Locus is called an allele. The ordered list of loci known for a particular genome is called a genetic map.
Nucleomorph	Nucleomorph s are small, reduced eukaryotic nuclei found in certain plastids. So far, only two groups of organisms are known to contain a Nucleomorph the cryptomonads of the supergroup Chromista and the chlorarachniophytes of the supergroup Rhizaria. The Nucleomorph s support the endosymbiotic theory, and are an evidence that the plastids of these organisms are so-called complex plastids.
Photosynthesis	Photosynthesis is a process that converts carbon dioxide into organic compounds, especially sugars, using the energy from sunlight. Photosynthesis occurs in plants, algae, and many species of Bacteria, but not in Archaea. Photosynthetic organisms are called photoautotrophs, since it allows them to create their own food.
Altruism	Altruism is a well-documented animal behaviour, which appears most obviously in kin relationships but may also be evident amongst wider social groups. In the science of ethology (the study of behavior), and more generally in the study of social evolution, Altruism refers to behavior by an individual that increases the fitness of another individual while decreasing the fitness of the actor. Research in evolutionary theory has been applied to social behaviour, including Altruism.
Adenosine	Adenosine is a nucleoside composed of a molecule of adenine attached to a ribose sugar molecule (ribofuranose) moiety via a β-N_9-glycosidic bond. Adenosine plays an important role in biochemical processes, such as energy transfer--as Adenosine triphosphate (ATP) and Adenosine diphosphate (ADP)--as well as in signal transduction as cyclic Adenosine monophosphate, cAMP. It is also an inhibitory neurotransmitter, believed to play a role in promoting sleep and suppressing arousal, with levels increasing with each hour an organism is awake. Adenosine is an endogenous purine nucleoside that modulates many physiological processes.
Deoxyadenosine diphosphate	Deoxyadenosine diphosphate is a derivative of the common nucleic acid ATP in which the -OH (hydroxyl) group on the 2" carbon on the nucleotide"s pentose has been removed (hence the deoxy- part of the name.) Additionally, the diphosphate of the name indicates that one of the phosphoryl groups of ATP has been removed, most likely by hydrolysis. Deoxyadenosine diphosphate would be abbreviated dADP.
Sulfolobus	In taxonomy, Sulfolobus is a genus of the Sulfolobaceae.

Sulfolobus species grow in volcanic springs with optimal growth occurring at pH 2-3 and temperatures of 75-80 °C, making them acidophiles and thermophiles respectively. Sulfolobus cells are irregularly shaped and flagellar.

Thermoplasma — In taxonomy, Thermoplasma is a genus of the Thermoplasma taceae.

Adaptation — In ocular physiology, Adaptation is the ability of the eye to adjust to various levels of darkness and light.

The human eye can function from very dark to very bright levels of light -- its sensing capabilities reach across nine orders of magnitude. This means that the brightest and the darkest light signal that the eye can sense are a factor of roughly one thousand million apart.

Erwinia — Erwinia is a genus of Enterobacteriaceae bacteria containing mostly plant pathogenic species which was named for the first phytobacteriologist, Erwin Smith. It is a gram negative bacterium related to E.coli, Shigella, Salmonella and Yersinia. It is primarily a rod-shaped bacteria.

Erwinia carotovora — Erwinia carotovora is a bacterium of the family Enterobacteriaceae, which is a plant pathogen with a diverse host range. It causes soft rot and blackleg of potato as well as slime flux on many different tree species.

Dehalococcoides — Dehalococcoides is a genus of bacteria within class Dehalococcoidetes that obtain energy via the oxidation of hydrogen gas and subsequent reductive dehalogenation of halogenated organic compounds.

Dehalococcoides ethenogenes strain 195 was first described in 1997, and is noted for its potential use in the bioremediation of tetrachloroethene (PCE) and trichloroethene (TCE) contaminated ground water sites. High D. ethenogenes population numbers and electron substrate concentrations are necessary for effective bioremediation of dense non-aqueous phase liquids (DNAPL) contamination sites.

Prochlorococcus — Prochlorococcus is a genus of very small (0.6 Âµm) marine cyanobacteria with an unusual pigmentation (chlorophyll b) belonging to photosynthetic picoplankton. It is probably the most abundant photosynthetic organism on Earth.

Although there had been several earlier records of very small chlorophyll-b-containing cyanobacteria in the ocean, Prochlorococcus was actually discovered in 1986 by Sallie W. (Penny) Chisholm of the Massachusetts Institute of Technology, Robert J. Olson of the Woods Hole Oceanographic Institution, and other collaborators in the Sargasso Sea using flow cytometry.

Pseudomonas — Pseudomonas is a genus of gamma proteobacteria, belonging to the larger family of pseudomonads.

Recently, 16S rRNA sequence analysis has redefined the taxonomy of many bacterial species. As a result the genus Pseudomonas includes strains formerly classified in the genera Chryseomonas and Flavimonas.

Regulatory sequence

A Regulatory sequence is a segment of DNA where regulatory proteins such as transcription factors bind preferentially. These regulatory proteins bind to short stretches of DNA called regulatory regions, which are appropriately positioned in the genome, usually a short distance "upstream" of the gene being regulated. By doing so, these regulatory proteins can recruit another protein complex, called the RNA polymerase.

Syringa

Syringa is a genus of about 20-25 species of flowering plants in the olive family (Oleaceae), native to Europe and Asia. Syringa josikaea

They are deciduous shrubs or small trees, ranging in size from 2-10 m tall, with stems up to 20-30 cm diameter. The leaves are opposite (occasionally in whorls of three) in arrangement, and their shape is simple and heart-shaped to broad lanceolate in most species, but pinnate in a few species (e.g. S. protolaciniata, S. pinnatifolia.)

Gene expression

Gene expression is the process by which information from a gene is used in the synthesis of a functional gene product. These products are often proteins, but in non-protein coding genes such as rRNA genes or tRNA genes, the product is a functional RNA.

Several steps in the Gene expression process may be modulated, including the transcription, RNA splicing, translation, and post-translational modification of a protein. Gene regulation gives the cell control over structure and function, and is the basis for cellular differentiation, morphogenesis and the versatility and adaptability of any organism.

Phenotype

A Phenotype is any observable characteristic or trait of an organism: such as its morphology, development, biochemical or physiological properties, or behavior. Phenotype s result from the expression of an organism"s genes as well as the influence of environmental factors and possible interactions between the two. The genotype of an organism is the inherited instructions it carries within its genetic code.

Blending inheritance

In Darwin"s time, biologists held to the theory of Blending inheritance--an offspring was an average of its parent. If an individual had one short parent and one tall parent, it would always be of some interim height between that of its parents. In turn, this interim height would become a theoretical extreme (either tall or short) that, in turn, bounded the potential height of its own offspring.

Promoter

In genetics, a Promoter is a region of DNA that facilitates the transcription of a particular gene. Promoter s are typically located near the genes they regulate, on the same strand and upstream (towards the 5" region of the sense strand.)

In order for transcription to take place, the enzyme that synthesizes RNA, known as RNA polymerase, must attach to the DNA near a gene.

Diatoms	Diatoms = "to cut", i.e., "cut in half") are a major group of eukaryotic algae, and are one of the most common types of phytoplankton. Most Diatoms are unicellular, although they can exist as colonies in the shape of filaments or ribbons (e.g. Fragillaria), fans (Meridion), zigzags (Tabellaria), or stellate colonies (Asterionella.) Diatoms are producers within the food chain.
Urea	Urea or carbamide is an organic compound with the chemical formula $(NH_2)_2CO$. The molecule has two amine ($-NH_2$) residues joined by a carbonyl (-CO-) functional group. Urea plays an important role in the metabolism of nitrogen-containing compounds by animals, and is the main nitrogen-containing substance in the urine of mammals. Being solid, colourless, odorless, neither acidic nor basic, highly soluble in water, and relatively non-toxic, Urea is widely used in fertilizers as a convenient source of nitrogen.
Urea cycle	The Urea cycle is a cycle of biochemical reactions occurring in many animals that produces urea $(NH_2)_2CO$ from ammonia (NH_3.) This cycle was the first metabolic cycle discovered (Hans Krebs and Kurt Henseleit, 1932.) In mammals, the Urea cycle takes place only in the liver.
Sequencing	In genetics and biochemistry, Sequencing means to determine the primary structure (or primary sequence) of an unbranched biopolymer. Sequencing results in a symbolic linear depiction known as a sequence which succinctly summarizes much of the atomic-level structure of the sequenced molecule. DNA Sequencing is the process of determining the nucleotide order of a given DNA fragment.
Functional genomics	Functional genomics is a field of molecular biology that attempts to make use of the vast wealth of data produced by genomic projects (such as genome sequencing projects) to describe gene (and protein) functions and interactions. Unlike genomics and proteomics, Functional genomics focuses on the dynamic aspects such as gene transcription, translation, and protein-protein interactions, as opposed to the static aspects of the genomic information such as DNA sequence or structures. Functional genomics includes function-related aspects of the genome itself such as mutation and polymorphism (such as SNP) analysis, as well as measurement of molecular activities.

Species	In biology, a Species is: · a taxonomic rank (the basic rank of biological classification) or · a unit at that rank There are many definitions of what kind of unit a Species is (or should be.) A common definition is that of a group of organisms capable of interbreeding and producing fertile offspring, and separated from other such groups with which interbreeding does not (normally) happen. Other definitions may focus on similarity of DNA or morphology. Some Species are further subdivided into sub Species , and here also there is no close agreement on the criteria to be used.
Fly	Nematocera (includes Eudiptera)Brachycera True flies are insects of the order Diptera , possessing a single pair of wings on the mesothorax and a pair of halteres, derived from the hind wings, on the metathorax. The presence of a single pair of wings distinguishes true flies from other insects with "Fly" in their name, such as mayflies, dragonflies, damselflies, stoneflies, whiteflies, fireflies, alderflies, dobsonflies, snakeflies, sawflies, caddisflies, butterflies or scorpionflies. Some true flies have become secondarily wingless, especially in the superfamily Hippoboscoidea, or among those that are inquilines in social insect colonies.
Maggot	Maggot is the common name of the larval phase of development in insects of the order Diptera (flies.) Sometimes the word is used to denote the larval stage of any insects. Anglers use Maggot s usually provided by commercial suppliers to catch non-predatory fish.
Endangered	An endangered species is "a species at risk of extinction because of human activity, changes in climate, changes in predator-prey ratios, etc., especially when officially designated as such by a governmental agency such as the U.S. Fish and Wildlife Service". A spider is a member of the order Araneae, one of many orders within the class arachnida, meaning it has a body consisting of two segments, a cephalothorax and an abdomen, as well as eight legs. All spiders are predators that feed off insects, and some larger species have been known to catch small reptiles, birds, and amphibians (Levi and Levi, 2002.)Until recently, insects and arachnids have not been considered for inclusion on threatened and endangered species lists.
Genetic	Genetics is the study of how living things receive common traits from previous generations. These traits are described by the Genetic information carried by a molecule called DNA. The instructions for constructing and operating an organism are contained in the organism"s DNA. Every living thing on earth has DNA in its cells. Genes are the hereditary components of DNA that occupy spots on chromosomes and determine characteristics in an organism.
Antibiotic resistance	Antibiotic resistance is the ability of a microorganism to withstand the effects of antibiotics. It is a specific type of drug resistance. antibiotic resistance evolves via natural selection acting upon random mutation, but it can also be engineered by applying an evolutionary stress on a population.

Genetic recombination	Genetic recombination is the process by which a strand of genetic material (usually DNA; but can also be RNA) is broken and then joined to a different DNA molecule. In eukaryotes recombination commonly occurs during meiosis as chromosomal crossover between paired chromosomes. This process leads to offspring having different combinations of genes from their parents and can produce new chimeric alleles.
Bacteria	The Bacteria are a large group of unicellular microorganisms. Typically a few micrometres in length, Bacteria have a wide range of shapes, ranging from spheres to rods and spirals. Bacteria are ubiquitous in every habitat on Earth, growing in soil, acidic hot springs, radioactive waste, water, and deep in the Earth"s crust, as well as in organic matter and the live bodies of plants and animals.
Horse murders	The Horse murders scandal was a form of insurance fraud in the United States in which expensive horses, many of them show jumpers, were insured against death, accident and then killed to collect the insurance money. It is not known how many horses were murdered between the mid 1970s and the mid-1990s, when a Federal Bureau of Investigation (FBI) investigation brought the horse killings to light, but the number is thought to be well over 50, and may have been as high as 100. In addition, in 1977, the heiress Helen Brach disappeared and was presumed by law enforcement agents to have been murdered by the perpetrators of these crimes, because she threatened to report their criminal activity to authorities; continuing investigations into Brach"s death began to uncover the insurance fraud in the 1990s.
Animals	Animals are a major group of mostly multicellular, eukaryotic organisms of the kingdom Animalia or Metazoa. Their body plan eventually becomes fixed as they develop, although some undergo a process of metamorphosis later on in their life. Most Animals are motile, meaning they can move spontaneously and independently.
African elephants	African elephants are the species of elephants in the genus Loxodonta, one of the two existing genera in Elephantidae. Although it is commonly believed that the genus was named by Georges Cuvier in 1825, Cuvier spelled it Loxodonte. An anonymous author romanized the spelling to Loxodonta and the ICZN recognizes this as the proper authority.
Elephant	Elephant s are large land mammals of the order Proboscidea and the family Elephant idae. There are three living species: the African Bush Elephant the African Forest Elephant and the Asian Elephant (also known as the Indian Elephant) Other species have become extinct since the last ice age, the Mammoths, dwarf forms of which may have survived as late as 2,000 BC, being the best-known of these.
Gene	A Gene is the basic unit of heredity in a living organism. All living things depend on Gene s. Gene s hold the information to build and maintain their cells and pass Gene tic traits to offspring.
Gene flow	In population genetics, gene flow is the transfer of alleles of genes from one population to another.

Migration into or out of a population may be responsible for a marked change in allele frequencies Immigration may also result in the addition of new genetic variants to the established gene pool of a particular species or population.

Genetic variation

Genetic variation, variation in alleles of genes, occurs both within and among populations. genetic variation is important because it provides the "raw material" for natural selection.

genetic variation among individuals within a population can be identified at a variety of levels.

Chromosome

A Chromosome is an organized structure of DNA and protein that is found in cells. It is a single piece of coiled DNA containing many genes, regulatory elements and other nucleotide sequences. Chromosome s also contain DNA-bound proteins, which serve to package the DNA and control its functions.

Drosophila

Drosophila has long been a favorite model system for geneticists and developmental biologists studying embryogenesis. The small size, short generation time, and large brood size makes it ideal for genetic studies. Transparent embryos facilitate developmental studies.

Polyploidy

Polyploidy occurs in cells and organisms when there are more than two paired (homologous) sets of chromosomes. Known paleo Polyploidy in eukaryotes

Most organisms are normally diploid, meaning they have two sets of chromosomes -- one set inherited from each parent. Polyploidy may occur due to abnormal cell division.

Fern

A Fern is any one of a group of about 20,000 species of plants classified in the phylum or division Pteridophyta, also known as Filicophyta. The group is also referred to as Polypodiophyta, or Polypodiopsida when treated as a subdivision of tracheophyta (vascular plants.) The term "pteridophyte" has traditionally been used to describe all seedless vascular plants, making it synonymous with " Fern s and Fern allies".

Genetic drift

Genetic drift or allelic drift is the change in the relative frequency with which a gene variant (allele) occurs in a population that results from the fact that alleles in offspring are a random sample of those in the parents, and because of the role of chance in determining whether a given individual survives and reproduces. A population"s allele frequency is the fraction of the gene copies that share a particular form.

Genetic drift is one of several evolutionary processes which lead to changes in allele frequencies over time.

Adaptation

In ocular physiology, Adaptation is the ability of the eye to adjust to various levels of darkness and light.

The human eye can function from very dark to very bright levels of light -- its sensing capabilities reach across nine orders of magnitude. This means that the brightest and the darkest light signal that the eye can sense are a factor of roughly one thousand million apart.

DNA	Deoxyribonucleic acid (DNA) is a nucleic acid that contains the genetic instructions used in the development and functioning of all known living organisms and some viruses. The main role of DNA molecules is the long-term storage of information. DNA is often compared to a set of blueprints or a recipe, or a code, since it contains the instructions needed to construct other components of cells, such as proteins and RNA molecules.
DNA sequence	A DNA sequence or genetic sequence is a succession of letters representing the primary structure of a real or hypothetical DNA molecule or strand, with the capacity to carry information as described by the central dogma of molecular biology. The possible letters are A, C, G, and T, representing the four nucleotide bases of a DNA strand -- adenine, cytosine, guanine, thymine -- covalently linked to a phosphodiester backbone. In the typical case, the sequences are printed abutting one another without gaps, as in the sequence AAAGTCTGAC, read left to right in the 5" to 3" direction.
Population	In biology, a population is the collection of inter-breeding organisms of a particular species; in sociology, a collection of human beings. Individuals within a population share a factor may be reduced by statistical means, but such a generalization may be too vague to imply anything. Demography is used extensively in marketing, which relates to economic units, such as retailers, to potential customers.
Rosemary	Rosemary is a woody, perennial herb with fragrant evergreen needle-like leaves. It is native to the Mediterranean region. It is a member of the mint family Lamiaceae, which also includes many other herbs.
Genetic diversity	Genetic diversity is a level of biodiversity that refers to the total number of genetic characteristics in the genetic makeup of a species. It is distinguished from genetic variability, which describes the tendency of genetic characteristics to vary. The academic field of population genetics includes several hypotheses and theories regarding genetic diversity.
Size	Size has been one of the most interesting aspects of cephalopod science to the general public Extinct taxa are also included.
Plants	Plants are living organisms belonging to the kingdom Plantae. They include familiar organisms such as trees, herbs, bushes, grasses, vines, ferns, mosses, and green algae. About 350,000 species of Plants, defined as seed Plants, bryophytes, ferns and fern allies, are estimated to exist currently.
Allele	An Allele) is one of a series of different forms of a gene. The word is a short from of allelomorph , which was used in the early days of genetics to describe variant forms of a gene detected as different phenotypes. Allele s are now understood to be alternative DNA sequences at the same physical gene locus, which may or may not result in different phenotypic traits.

Allele frequency

Allele frequency is the proportion of all variants (alleles) that is made up of a particular gene variant. In other words, it is the number of copies of a particular allele divided by the number of copies of all alleles at the genetic place (locus) in a population. It can be expressed for example as a percentage.

Mutation

In biology, mutation s are changes to the nucleotide sequence of the genetic material of an organism. mutation s can be caused by copying errors in the genetic material during cell division, by exposure to ultraviolet or ionizing radiation, chemical mutagens, or viruses, or can be induced by the organism itself, by cellular processes such as hyper mutation . In multicellular organisms with dedicated reproductive cells, mutation s can be subdivided into germ line mutation s, which can be passed on to descendants through the reproductive cells, and somatic mutation s, which involve cells outside the dedicated reproductive group and which are not usually transmitted to descendants.

Drift Migration

Drift migration is the phenomenon in which migrating birds are blown off course by the winds at the time they are in flight. It is more likely to happen to birds heading south in autumn because the large numbers of inexperienced young birds are less able to compensate than the adults heading north in spring.

Drift is assisted by disorientating conditions such as mist or drizzle, and can result in large numbers of birds arriving together in an area in which they are not normally seen.

Lek

A Lek is a gathering of males, of certain animal species, for the purposes of competitive mating display. Leks assemble before and during the breeding season, on a daily basis. The same group of males meet at a traditional place and take up the same individual positions on an arena, each occupying and defending a small territory or court.

Hox genes

Hox genes are a group of related genes that specify the anterior-posterior axis and segment identity of metazoan organisms during early embryonic development. These genes are critical for the proper number and placement of embryonic segment structures (such as legs, antennae, and eyes.)

Homeotic genes are defined by a DNA sequence known as the homeobox, which is a sequence of 180 nucleotides that code for a protein domain known as the homeodomain.

Ectocarpene

Ectocarpene is a sexual attractant found with several species of brown algae (Phaeophyceae.) The substance has a fruity scent and can be sensed by humans when millions of algae gametes swarm the seawater and the females start emitting the substance to attract the male gametes.

Ectocarpene was the first isolated algal pheromone.

Mating

In biology, Mating is the pairing of opposite-sex or hermaphroditic organisms for copulation and, in social animals, also to raise their offspring. For animals, Mating methods include random Mating, disassortative Mating, assortative Mating, or a Mating pool.

In some birds, for example, it includes nest-building and feeding offspring.

Hybrid	Hybrid has two meanings. The first meaning is the result of interbreeding between two animals or plants of different taxa. Hybrids between different subspecies within a species (such as between the Bengal tiger and Siberian tiger) are known as intra-specific hybrids.
Key Habitat Site	A Key Habitat Site is a Canadian Wildlife Service designation for an area that supports at least 1% of the country"s population of any migratory bird species at any time. There may be overlap with areas designated as a migratory bird sanctuary or national wildlife area.
Gill	A Gill is an anatomical structure found in many aquatic organisms. It is a respiration organ whose function is the extraction of oxygen from water and the excretion of carbon dioxide. The microscopic structure of a Gill is such that it presents a very large surface area to the external environment.

Charles Robert Darwin	Charles Robert Darwin FRS (12 February 1809 - 19 April 1882) was an English naturalist who realised and presented compelling evidence that all species of life have evolved over time from common ancestors, through the process he called natural selection. The fact that evolution occurs became accepted by the scientific community and much of the general public in his lifetime, while his theory of natural selection came to be widely seen as the primary explanation of the process of evolution in the 1930s, and now forms the basis of modern evolutionary theory. In modified form, Darwin"s scientific discovery is the unifying theory of the life sciences, providing logical explanation for the diversity of life.
Leukemia inhibitory factor	Leukemia inhibitory factor an interleukin 6 class cytokine, is a chemical in cells that affects their growth and development. Leukemia inhibitory factor derives its name from its ability to induce the terminal differentiation of myeloid leukaemic cells. Other properties attributed to the cytokine include: the growth promotion and cell differentiation of different types of target cells, influence on bone metabolism, cachexia, neural development, embryogenesis and inflammation.
Species	In biology, a Species is: · a taxonomic rank (the basic rank of biological classification) or · a unit at that rank There are many definitions of what kind of unit a Species is (or should be.) A common definition is that of a group of organisms capable of interbreeding and producing fertile offspring, and separated from other such groups with which interbreeding does not (normally) happen. Other definitions may focus on similarity of DNA or morphology. Some Species are further subdivided into sub Species , and here also there is no close agreement on the criteria to be used.
RNA	Ribonucleic acid (RNA) is a biologically important type of molecule that consists of a long chain of nucleotide units. Each nucleotide consists of a nitrogenous base, a ribose sugar, and a phosphate. RNA is very similar to DNA, but differs in a few important structural details: in the cell, RNA is usually single-stranded, while DNA is usually double-stranded; RNA nucleotides contain ribose while DNA contains deoxyribose (a type of ribose that lacks one oxygen atom); and RNA has the base uracil rather than thymine that is present in DNA. RNA is transcribed from DNA by enzymes called RNA polymerases and is generally further processed by other enzymes.
RNA World hypothesis	The RNA world hypothesis proposes that a world filled with life based on ribonucleic acid (RNA) predates the current world of life based on deoxyribonucleic acid (DNA.) RNA, which can both store information like DNA and act as an enzyme, may have supported cellular or pre-cellular life. Some hypotheses as to the origin of life present RNA-based catalysis and information storage as the first step in the evolution of cellular life.

Form	In botanical nomenclature, a Form is a low-level taxonomic rank below that of variety; it is an infraspecific taxon. Its name consists of three parts: a genus name, a specific epithet, and an infraspecific epithet. The abbreviation "f." or the full "forma" should be put before the infraspecific epithet to indicate the rank.
Cellular life	Cellular life is life with cells. The usual meaning of life as we know it expects that an organism have either unicellular or multicellular structure. But implied by the modifier "cellular" is the existence of earth-based life that is non-cellular.
DNA	Deoxyribonucleic acid (DNA) is a nucleic acid that contains the genetic instructions used in the development and functioning of all known living organisms and some viruses. The main role of DNA molecules is the long-term storage of information. DNA is often compared to a set of blueprints or a recipe, or a code, since it contains the instructions needed to construct other components of cells, such as proteins and RNA molecules.
Protein	Protein s are organic compounds made of amino acids arranged in a linear chain. The amino acids in a polymer chain are joined together by the peptide bonds between the carboxyl and amino groups of adjacent amino acid residues. The sequence of amino acids in a protein is defined by the sequence of a gene, which is encoded in the genetic code.
Ribonucleotide	A Ribonucleotide is a nucleotide in which a purine or pyrimidine base is linked to a ribose molecule. The base may be adenine (A), guanine (G), cytosine (C), or uracil (U.) Note that thymine (T), which is found in deoxy Ribonucleotide s, is not found as a Ribonucleotide in living beings.
Ribozyme	A Ribozyme is an RNA molecule that catalyzes a chemical reaction. Many natural Ribozyme s catalyze either the hydrolysis of one of their own phosphodiester bonds, or the hydrolysis of bonds in other RNAs, but they have also been found to catalyze the aminotransferase activity of the ribosome. Investigators studying the origin of life have produced Ribozyme s in the laboratory that are capable of catalyzing their own synthesis under very specific conditions, such as an RNA polymerase Ribozyme
Enzymes	Enzymes are biomolecules that catalyze (i.e., increase the rates of) chemical reactions. Nearly all known Enzymes are proteins. However, certain RNA molecules can be effective biocatalysts too.
Phenotype	A Phenotype is any observable characteristic or trait of an organism: such as its morphology, development, biochemical or physiological properties, or behavior. Phenotype s result from the expression of an organism"s genes as well as the influence of environmental factors and possible interactions between the two. The genotype of an organism is the inherited instructions it carries within its genetic code.

Intron	An Intron is a DNA region within a gene that is not translated into protein. These non-coding sections are transcribed to precursor mRNA (pre-mRNA) and some other RNAs (such as long noncoding RNAs), and subsequently removed by a process called splicing during the processing to mature RNA. After Intron splicing (ie. removal), the mRNA consists only of exon derived sequences, which are translated into a protein.
Genome	In classical genetics, the Genome of a diploid organism including eukarya refers to a full set of chromosomes or genes in a gamete; thereby, a regular somatic cell contains two full sets of Genome s. In haploid organisms, including bacteria, archaea, viruses, and mitochondria, a cell contains only a single set of the Genome usually in a single circular or contiguous linear DNA (or RNA for retroviruses.) In modern molecular biology the Genome of an organism is its hereditary information encoded in DNA (or, for retroviruses, RNA.)
Gene	A Gene is the basic unit of heredity in a living organism. All living things depend on Gene s. Gene s hold the information to build and maintain their cells and pass Gene tic traits to offspring.
Ribosome	Ribosome s are complexes of RNA and protein that are found in all cells with nuclei. Ribosome s from bacteria, archaea and eukaryotes (the three domains of life on Earth), have significantly different structure and RNA. The Ribosome s in the mitochondria of eukaryotic cells resemble those in bacteria, reflecting the evolutionary origin of this organelle. The Ribosome functions in the expression of the genetic code from nucleic acid into protein, in a process called translation.
Blending inheritance	In Darwin"s time, biologists held to the theory of Blending inheritance--an offspring was an average of its parent. If an individual had one short parent and one tall parent, it would always be of some interim height between that of its parents. In turn, this interim height would become a theoretical extreme (either tall or short) that, in turn, bounded the potential height of its own offspring.
Chromosome	A Chromosome is an organized structure of DNA and protein that is found in cells. It is a single piece of coiled DNA containing many genes, regulatory elements and other nucleotide sequences. Chromosome s also contain DNA-bound proteins, which serve to package the DNA and control its functions.
Inversion	An inversion is a chromosome rearrangement in which a segment of a chromosome is reversed end to end. An inversion occurs when a single chromosome undergoes breakage and rearrangement within itself. Inversions are of two types: paracentric and pericentric.
Oligonucleotide	An Oligonucleotide is a short nucleic acid polymer, typically with twenty or fewer bases. Although they can be formed by bond cleavage of longer segments, they are now more commonly synthesized by polymerizing individual nucleotide precursors. Automated synthesizers allow the synthesis of Oligonucleotide s up to 160 to 200 bases.

Reverse transcriptase

In biochemistry, a Reverse transcriptase is a DNA polymerase enzyme that transcribes single-stranded RNA into single-stranded DNA. It also helps in the formation of a double helix DNA once the RNA has been reverse transcribed into a single strand cDNA. Normal transcription involves the synthesis of RNA from DNA; hence, reverse transcription is the reverse of this.

Well studied Reverse transcriptase s include:

· HIV-1 Reverse transcriptase from human immunodeficiency virus type 1
· M-MLV Reverse transcriptase from the Moloney murine leukemia virus
· AMV Reverse transcriptase from the avian myeloblastosis virus
· Telomerase Reverse transcriptase that maintains the telomeres of eukaryotic chromosomes

The enzyme is encoded and used by reverse-transcribing viruses, which use the enzyme during the process of replication. Reverse-transcribing RNA viruses, such as retroviruses, use the enzyme to reverse-transcribe their RNA genomes into DNA, which is then integrated into the host genome and replicated along with it.

Aptamers

Aptamers are oligonucleic acid or peptide molecules that bind to a specific target molecule. Aptamers are usually created by selecting them from a large random sequence pool, but natural Aptamers also exist in riboswitches. Aptamers can be used for both basic research and clinical purposes as macromolecular drugs.

Population

In biology, a population is the collection of inter-breeding organisms of a particular species; in sociology, a collection of human beings. Individuals within a population share a factor may be reduced by statistical means, but such a generalization may be too vague to imply anything. Demography is used extensively in marketing, which relates to economic units, such as retailers, to potential customers.

Amino acid

In chemistry, an Amino acid is a molecule containing both amine and carboxyl functional groups. These molecules are particularly important in biochemistry, where this term refers to alpha- Amino acid s with the general formula $H_2NCHRCOOH$, where R is an organic substituent. In the alpha Amino acid s, the amino and carboxylate groups are attached to the same carbon atom, which is called the α-carbon.

Nucleic acid

A Nucleic acid is a macromolecule composed of chains of monomeric nucleotides. In biochemistry these molecules carry genetic information or form structures within cells. The most common Nucleic acid s are deoxyribo Nucleic acid (D Nucleic acid) and ribo Nucleic acid (R Nucleic acid .)

Precursor

In chemistry, a Precursor is a compound that participates in the chemical reaction that produces another compound. In biochemistry, the term "Precursor" is used more specifically to refer to a chemical compound preceding another in a metabolic pathway.

David	DAVID (the Database for Annotation, Visualization and Integrated Discovery) is a free online bioinformatics resources developed by the Laboratory of Immunopathogenesis and Bioinformatics (LIB.) All tools in the DAVID Bioinformatics Resources aim to provide functional interpretation of large lists of genes derived from genomic studies, e.g. microarray and proteomics studies. DAVID can be found at http://DAVID.niaid.nih.gov or http://DAVID.abcc.ncifcrf.gov The newly updated DAVID Bioinformatics Resources consists of the DAVID Knowledgebase and five integrated, web-based functional annotation tool suites: the DAVID Gene Functional Classification Tool, the DAVID Functional Annotation Tool, the DAVID Gene ID Conversion Tool, the DAVID Gene Name Viewer and the DAVID NIAID Pathogen Genome Browser.
Iron	Iron is a chemical element with the symbol Fe and atomic number 26. Iron is a group 8 and period 4 element. Iron and Iron alloys (steels) are by far the most common metals and the most common ferromagnetic materials in everyday use.
Polycyclic aromatic hydrocarbons	Polycyclic aromatic hydrocarbons (Polycyclic aromatic hydrocarbons s) are chemical compounds that consist of fused aromatic rings and do not contain heteroatoms or carry substituents. Polycyclic aromatic hydrocarbons s occur in oil, coal, and tar deposits, and are produced as byproducts of fuel burning (whether fossil fuel or biomass.) As a pollutant, they are of concern because some compounds have been identified as carcinogenic, mutagenic, and teratogenic.
Bacteria	The Bacteria are a large group of unicellular microorganisms. Typically a few micrometres in length, Bacteria have a wide range of shapes, ranging from spheres to rods and spirals. Bacteria are ubiquitous in every habitat on Earth, growing in soil, acidic hot springs, radioactive waste, water, and deep in the Earth''s crust, as well as in organic matter and the live bodies of plants and animals.
Bacillus	Bacillus is a genus of rod-shaped bacteria and a member of the division Firmicutes. Bacillus species are either obligate or facultative anaerobes, and test positive for the enzyme catalase. Ubiquitous in nature, Bacillus includes both free-living and pathogenic species.
Bacillus subtilis	Bacillus subtilis is a Gram-positive, catalase-positive bacterium commonly found in soil. A member of the genus Bacillus, B. subtilis is rod-shaped, and has the ability to form a tough, protective endospore, allowing the organism to tolerate extreme environmental conditions. Unlike several other well-known species, B. subtilis has historically been classified as an obligate aerobe, though recent research has demonstrated that this is not strictly correct.

Radiation

In physics, radiation describes any process in which energy emitted by one body travels through a medium or through space, ultimately to be absorbed by another body. Non-physicists often associate the word with ionizing radiation, but it can also refer to electromagnetic radiation (i.e., radio waves, infrared light, visible light, ultraviolet light, and X-rays) which can also be ionizing radiation, to acoustic radiation, or to other more obscure processes. What makes it radiation is that the energy radiates (i.e., it travels outward in straight lines in all directions) from the source.

Spore

In biology, a Spore is a reproductive structure that is adapted for dispersal and surviving for extended periods of time in unfavorable conditions. Spore s form part of the life cycles of many bacteria, plants, algae, fungi and some protozoans. A chief difference between Spore s and seeds as dispersal units is that Spore s have very little stored food resources compared with seeds.

Edward

The Edward mango is a monoembryonic mango cultivar grown predominantly in Florida. It is considered by many to be among the finest tasting mangoes in the world; however, its poor yields have restrained the Edward from developing into a commercially significant variety.

The Edward was first propagated in the 1920s by Edward Simmonds of the Plant Introduction Garden in Miami, Florida and is believed to be a hybrid cross of Haden and Carabao mango cultivars.

Sleep

Sleep is a natural state of bodily rest observed in humans and other animals. It is distinguished from quiet wakefulness by a decreased ability to react to stimuli, and it is more easily reversible than hibernation or coma. It is common to all mammals and birds, and is also seen in many reptiles, amphibians, and fish.

Alanine

Alanine is an α-amino acid with the chemical formula $CH_3CH(NH_2)COOH$. The L-isomer is one of the 20 proteinogenic amino acids, i.e. the building blocks of proteins. Its codons are GCU, GCC, GCA, and GCG. It is classified as a nonpolar amino acid. L-Alanine is second only to leucine, accounting for 7.8% of the primary structure in a sample of 1,150 proteins.

Glycin

Glycin is N-substituted p-aminophenol. It is a photographic developing agent used in classic B'W developer solutions. It is a derivative of the amino acid Glycin e.

Glycine

Glycine is the organic compound with the formula NH_2CH_2COOH. It is the smallest of the 20 amino acids commonly found in proteins, coded by codons GGU, GGC, GGA and GGG. Glycine is unique among the proteinogenic amino acids in that it is not chiral. Most proteins incorporate only small quantities of Glycine. A notable exception is collagen, which contains about 35% Glycine.

Miller

A Miller usually refers to a person who operates a mill, a machine to grind a cereal crop to make flour.

Milling is among the oldest of human occupations. "Miller", "Milne" and other variants are common surnames, as are their equivalents in other languages around the world .

Prebiotics

Prebiotics are non-digestible food ingredients that stimulate the growth or activity of bacteria in the digestive system which are beneficial to the health of the body. They are considered a functional food.

Typically, Prebiotics are carbohydrates (such as oligosaccharides), but the definition does not preclude non-carbohydrates.

Nucleotide

Nucleotide s are molecules that, when joined together, make up the structural units of RNA and DNA. Additionally, Nucleotide s play central roles in metabolism. In that capacity, they serve as sources of chemical energy (adenosine triphosphate and guanosine triphosphate), participate in cellular signaling (cyclic guanosine monophosphate and cyclic adenosine monophosphate), and are incorporated into important cofactors of enzymatic reactions (coenzyme A, flavin adenine di Nucleotide , flavin mono Nucleotide , and nicotinamide adenine di Nucleotide phosphate.) Figure 1: Structural elements of the most common Nucleotide s Figure 2: Ribose structure indicating numbering of carbon atoms

A Nucleotide is composed of a nucleobase (nitrogenous base) and a five-carbon sugar (either ribose or 2"-deoxyribose), and one to three phosphate groups.

Pyrimidine

Pyrimidine is a heterocyclic aromatic organic compound similar to benzene and pyridine, containing two nitrogen atoms at positions 1 and 3 of the six-member ring. It is isomeric with two other forms of diazine.

Three nucleobases found in nucleic acids (cytosine, thymine, and uracil) are Pyrimidine derivatives:

In DNA and RNA, these bases form hydrogen bonds with their complementary purines.

Ribose

Ribose, primarily occurring as D-Ribose, is an organic compound that occurs widely in nature. It is an aldopentose, that is a monosaccharide containing five carbon atoms that, in its acyclic form, has an aldehyde functional group at one end. Typically, this species exists in the cyclic form, as depicted in the upper right.

Sugar

Sugar is a class of edible crystalline substances, mainly sucrose, lactose, and fructose. Human taste buds interpret its flavor as sweet. Sugar as a basic food carbohydrate primarily comes from Sugar cane and from Sugar beet, but also appears in fruit, honey, sorghum, Sugar maple (in maple syrup), and in many other sources.

Primer

A Primer is a strand of nucleic acid that serves as a starting point for DNA replication. They are required because the enzymes that catalyze replication, DNA polymerases, can only add new nucleotides to an existing strand of DNA. The polymerase starts replication at the 3"-end of the Primer, and copies the opposite strand.

In most cases of natural DNA replication, the Primer for DNA synthesis and replication is a short strand of RNA (which can be made de novo.)

Apex

Apex is an anatomical term for the tip of the mollusk shell of a gastropod or scaphopod mollusk.

The word "Apex" is most often used to mean the tip of the spire of the shell of a gastropod. The Apex is the first-formed, and therefore the oldest, part of the shell.

Cyanobacteria

Cyanobacteria blue-green bacteria or Cyanophyta, is a phylum of bacteria that obtain their energy through photosynthesis. The name "Cyanobacteria" comes from the color of the bacteria . They are a significant component of the marine nitrogen cycle and an important primary producer in many areas of the ocean, but are also found in habitats other than the marine environment; in particular Cyanobacteria are known to occur in both freshwater, hypersaline inland lakes and in arid areas where they are a major component of biological soil crusts.

Ribosomal RNA

Ribosomal RNA is the central component of the ribosome, the protein manufacturing machinery of all living cells. The function of the rRNA is to provide a mechanism for decoding mRNA into amino acids and to interact with the tRNAs during translation by providing peptidyl transferase activity.The tRNA then brings the necessary amino acids corresponding to the appropriate mRNA codon.

The ribosome is composed of two subunits P, and E.

· The A site in the ribosome binds to an aminoacyl-tRNA (a tRNA bound to an amino acid.)
· The amino (NH2) group of the aminoacyl-tRNA, which contains the new amino acid, attacks the ester linkage of peptidyl-tRNA (contained within the P site), which contains the last amino acid of the growing chain, forming a new peptide bond. This reaction is catalyzed by peptidyl transferase.
· The tRNA that was holding on the last amino acid is moved to the E site, and what used to be the aminoacyl-tRNA is now the peptidyl-tRNA.

A single mRNA can be translated simultaneously by multiple ribosomes.

Both prokaryotic and eukaryotic can be broken down into two subunits (the S in 16S represents Svedberg units):

Note that the S units of the subunits cannot simply be added because they represent measures of sedimentation rate rather than of mass.

Genetic

Genetics is the study of how living things receive common traits from previous generations. These traits are described by the Genetic information carried by a molecule called DNA. The instructions for constructing and operating an organism are contained in the organism"s DNA. Every living thing on earth has DNA in its cells. Genes are the hereditary components of DNA that occupy spots on chromosomes and determine characteristics in an organism.

Sequencing

In genetics and biochemistry, Sequencing means to determine the primary structure (or primary sequence) of an unbranched biopolymer. Sequencing results in a symbolic linear depiction known as a sequence which succinctly summarizes much of the atomic-level structure of the sequenced molecule.

DNA Sequencing is the process of determining the nucleotide order of a given DNA fragment.

Animals	Animals are a major group of mostly multicellular, eukaryotic organisms of the kingdom Animalia or Metazoa. Their body plan eventually becomes fixed as they develop, although some undergo a process of metamorphosis later on in their life. Most Animals are motile, meaning they can move spontaneously and independently.
Archaea	The Archaea are a group of single-celled microorganisms. A single individual or species from this domain is called an archaeon (sometimes spelled "archeon".) They have no cell nucleus or any other organelles within their cells.
Crenarchaeota	In taxonomy, the Crenarchaeota are a phylum of the Archaea. Initially, the Crenarchaeota were thought to be extremophiles (e.g., thermophilic and psychrophilic organisms) but recent studies have identified them as the most abundant archaea in the marine environment. Originally, they were separated from the other archaea based on rRNA sequences; since then physiological features, such as lack of histones have supported this division.
Euryarchaeota	In the taxonomy of microorganisms, the Euryarchaeota are a phylum of the Archaea. The Euryarchaeota include the methanogens, which produce methane and are often found in intestines, the halobacteria, which survive extreme concentrations of salt, and some extremely thermophilic aerobes and anaerobes. They are separated from the other archaeans based mainly on rRNA sequences.
Fungi	A fungus is a eukaryotic organism that is a member of the kingdom fungi . The fungi are a monophyletic group, also called the Eumycota , that is phylogenetically distinct from the structurally similar slime molds (myxomycetes) and water molds (oomycetes.) The fungi are heterotrophic organisms possessing a chitinous cell wall, with most species growing as multicellular filaments called hyphae forming a mycelium; some species also grow as single cells.
Haloarchaea	Haloarchaea are microrganisms member of the halophile community, in that they require high salt concentrations to grow. They are a distinct evolutionary branch of the Archaea, and are generally considered extremophiles, although not all members of this group can be considered as such. Salt ponds with pink colored Haloarchaea Haloarchaea require salt concentrations in excess of 2 M (or about 10%) to grow, and optimal growth usually occurs at much higher concentrations, typically 20-25%.
Mycobacterium	Mycobacterium is a genus of Actinobacteria, given its own family, the Mycobacteriaceae. The genus includes pathogens known to cause serious diseases in mammals, including tuberculosis and leprosy. The Latin prefix "myco--" means both fungus and wax; its use here relates to the "waxy" compounds in the cell wall.

Nostoc	Nostoc is a genus of fresh water cyanobacteria that forms spherical colonies composed of filaments of moniliform cells in a gelatinous sheath. When on the ground, a Nostoc colony is ordinarily not seen; but after a rain it swells up into a conspicuous jellylike mass, which was once thought to have fallen from the sky, hence the popular names, fallen star and star jelly. It is also called witches" butter .
Plants	Plants are living organisms belonging to the kingdom Plantae. They include familiar organisms such as trees, herbs, bushes, grasses, vines, ferns, mosses, and green algae. About 350,000 species of Plants, defined as seed Plants, bryophytes, ferns and fern allies, are estimated to exist currently.
Purple bacteria	Purple bacteria or purple photosynthetic bacteria are proteobacteria that are phototrophic, i.e. capable of producing energy through photosynthesis. They are pigmented with bacteriochlorophyll a or b, together with various carotenoids. These give them colours ranging between purple, red, brown, and orange.
Tuberculosis	Tuberculosis is a common and often deadly infectious disease caused by mycobacteria, in humans mainly Mycobacterium Tuberculosis . Tuberculosis usually attacks the lungs (as pulmonary TB) but can also affect the central nervous system, the lymphatic system, the circulatory system, the genitourinary system, the gastrointestinal system, bones, joints, and even the skin. Other mycobacteria such as Mycobacterium bovis, Mycobacterium africanum, Mycobacterium canetti, and Mycobacterium microti also cause Tuberculosis, but these species are less common in humans.
Methanococcus	In taxonomy, Methanococcus is a genus of the Methanococcaceae. Methanococcus is a genus of coccoid methanogens. They are all mesophiles, except the thermophilic M. thermolithotrophicus and the hyperthermophilic M. jannaschii.
Caenorhabditis	Caenorhabditis is a genus of nematodes which live in bacteria-rich environments like compost piles and decaying dead animals. It contains the noted model organism Caenorhabditis elegans and several other species for which a genome sequence is either available or currently being determined. The two most-studied species in this genus (C. elegans and C. briggsae) are both hermaphrodites whereas all other species are gonochoristic (they have male and female sexes.)
Korarchaeota	In taxonomy, the Korarchaeota are a phylum of the Archaea. The Korarchaeota are a group of Archaea that have been found only in high temperature hydrothermal environments. Analysis of their 16S rRNA gene sequences suggests that they are a deeply-branching lineage that does not belong to the main archaeal groups, Crenarchaeota and Euryarchaeota.

Antibiotic resistance	Antibiotic resistance is the ability of a microorganism to withstand the effects of antibiotics. It is a specific type of drug resistance. antibiotic resistance evolves via natural selection acting upon random mutation, but it can also be engineered by applying an evolutionary stress on a population.
Borrelia	Borrelia is a genus of bacteria of the spirochete phylum. It causes borreliosis, a zoonotic, vector-borne disease transmitted primarily by ticks and some by lice, depending on the species. There are 36 known species of Borrelia.
Thermotoga	Thermotoga are thermophile or hyperthermophile bacteria whose cell is wrapped in an outer "toga" membrane. They were named by microbiologist Karl Stetter. They metabolize carbohydrates.
Horse murders	The Horse murders scandal was a form of insurance fraud in the United States in which expensive horses, many of them show jumpers, were insured against death, accident and then killed to collect the insurance money. It is not known how many horses were murdered between the mid 1970s and the mid-1990s, when a Federal Bureau of Investigation (FBI) investigation brought the horse killings to light, but the number is thought to be well over 50, and may have been as high as 100. In addition, in 1977, the heiress Helen Brach disappeared and was presumed by law enforcement agents to have been murdered by the perpetrators of these crimes, because she threatened to report their criminal activity to authorities; continuing investigations into Brach"s death began to uncover the insurance fraud in the 1990s.
Blytt-Sernander	The Blytt-Sernander classification is a series of north European climatic periods or phases based on the study of Danish peat bogs by Axel Blytt (1876) and Rutger Sernander (1908.) The classification was incorporated into a sequence of pollen zones later defined by Lennart von Post, one of the founders of palynology. Layers in peat were first noticed by Heinrich Dau in 1829.
Root	In vascular plants, the Root is the organ of a plant that typically lies below the surface of the soil. This is not always the case, however, since a Root can also be aerial (growing above the ground) or aerating (growing up above the ground or especially above water.) Furthermore, a stem normally occurring below ground is not exceptional either
Gloeocapsa	Gloeocapsa) is a genus of photoautotrophic bacteria, (cyanobacteria), and is a prokaryote. The cells secrete individual gelatinous sheaths which can often be seen as sheaths around recently divided cells within outer sheaths. Recently divided cell pairs often appear to be only one cell since the new cells cohere temporarily.
Lyngbya	Lyngbya is a genus of cyanobacteria that is one of the causes of the human skin irritation called swimmer"s itch. This skin irritation is unrelated to cercarial dermatitis. Some Lyngbya species also cause destruction of aquatic ecosystems when they form dense floating mats in the water.

Spirulina	Spirulina is the common name for human and animal food supplements produced primarily from two species of cyanobacteria: Arthrospira platensis, and Arthrospira maxima. These and other Arthrospira species were once classified in the genus Spirulina. There is now agreement that they are a distinct genus, and that the food species belong to Arthrospira; nonetheless, the older term Spirulina remains the popular name.
Human	A Human is a member of a species of bipedal primates in the family Hominidae . DNA and fossil evidence indicates that modern Human s originated in east Africa about 200,000 years ago. When compared to other animals and primates, Human s have a highly developed brain, capable of abstract reasoning, language, introspection and problem solving.
Influenza	Influenza, commonly referred to as the flu, is an infectious disease caused by RNA viruses of the family Orthomyxoviridae (the Influenza viruses), that affects birds and mammals. The name Influenza comes from the Italian Influenza, meaning "influence" . The most common symptoms of the disease are chills, fever, sore throat, muscle pains, severe headache, coughing, weakness and general discomfort.
Phagocytosis	Phagocytosis is the cellular process of phagocytes and protists of engulfing solid particles by the cell membrane to form an internal phagosome, which is a food vacuole, or pteroid. Phagocytosis is a specific form of endocytosis involving the vesicular internalization of solid particles, such as bacteria, and is therefore distinct from other forms of endocytosis such as pinocytosis, the vesicular internalization of various liquids. Phagocytosis is involved in the acquisition of nutrients for some cells, and in the immune system it is a major mechanism used to remove pathogens and cell debris.
Virus	A virus is a microscopic infectious agent that can reproduce only inside a host cell. virus es infect all types of organisms: from animals and plants, to bacteria and archaea. Since the initial discovery of tobacco mosaic virus by Martinus Beijerinck in 1898, more than 5,000 types of virus have been described in detail, although most types of virus remain undiscovered.
RNA polymerase	RNA polymerase is an enzyme that produces RNA. In cells, RNAP is needed for constructing RNA chains from DNA genes as templates, a process called transcription. RNA polymerase enzymes are essential to life and are found in all organisms and many viruses. In chemical terms, RNAP is a nucleotidyl transferase that polymerizes ribonucleotides at the 3" end of an RNA transcript.

Scale	In most biological nomenclature, a Scale is a small rigid plate that grows out of an animal"s skin to provide protection. In lepidopteran species, scales are plates on the surface of the insect wing, and provide coloration. Scales are quite common and have evolved multiple times with varying structure and function.
Allele	An Allele) is one of a series of different forms of a gene. The word is a short from of allelomorph , which was used in the early days of genetics to describe variant forms of a gene detected as different phenotypes. Allele s are now understood to be alternative DNA sequences at the same physical gene locus, which may or may not result in different phenotypic traits.
Allele frequency	Allele frequency is the proportion of all variants (alleles) that is made up of a particular gene variant. In other words, it is the number of copies of a particular allele divided by the number of copies of all alleles at the genetic place (locus) in a population. It can be expressed for example as a percentage.
Arabidopsis thaliana	Arabidopsis thaliana , is a small flowering plant native to Europe, Asia, and northwestern Africa. A spring annual with a relatively short life cycle, Arabidopsis is popular as a model organism in plant biology and genetics. Its genome is one of the smallest plant genomes and was the first plant genome to be sequenced.
Escherichia	Escherichia is a genus of Gram-negative, non-spore forming, facultatively anaerobic, rod-shaped bacteria from the family Enterobacteriaceae. Inhabitants of the gastrointestinal tracts of warm-blooded animals, Escherichia species provide a portion of the microbially-derived vitamin K for their host. While many Escherichia are harmless commensals, particular strains of some species are human pathogens, and are known as the most common cause of urinary tract infections, significant sources of gastrointestinal disease, ranging from simple diarrhea to dysentery-like conditions, as well as a wide-range of other pathogenic states.
Escherichia coli	Escherichia coli , is a Gram negative bacterium that is commonly found in the lower intestine of warm-blooded organisms . Most E. coli strains are harmless, but some, such as serotype O157:H7, can cause serious food poisoning in humans, and are occasionally responsible for costly product recalls. The harmless strains are part of the normal flora of the gut, and can benefit their hosts by producing vitamin K_2, or by preventing the establishment of pathogenic bacteria within the intestine.
Nematode	The "roundworms" or " Nematode s" (phylum Nematoda) are the most diverse phylum of pseudocoelomates, and one of the most diverse of all animals. Nematode species are very difficult to distinguish; over 80,000 have been described, of which over 15,000 are parasitic. It has been estimated that the total number of described and undescribed roundworms might be more than 500,000.

Saccharomyces

Saccharomyces is a genus in the kingdom of fungi that includes many species of yeast. Saccharomyces is from Latin meaning sugar fungi. Many members of this genus are considered very important in food production.

Saccharomyces cerevisiae

Saccharomyces cerevisiae is a species of budding yeast. It is perhaps the most useful yeast owing to its use since ancient times in baking and brewing. It is believed that it was originally isolated from the skins of grapes

Mustard

Mustard is a thick yellowish-brown paste with a sharp taste made from the ground seeds of a Mustard plant (white or yellow Mustard, Sinapis hirta; brown or Indian Mustard, Brassica juncea; or black Mustard, Brassica nigra.) The ground Mustard seeds are mixed with water, vinegar or other liquids, and sometimes other flavorings and spices. A strong Mustard can cause the eyes to water, sting the palate and inflame the nasal passages.

Worm

The term Worm is used to describe many different distantly-related animals which have a long cylindrical body and no legs.

Most animals called " Worm s" are invertebrates, but the term is also used for the amphibian caecilians and the slow Worm Anguis, a legless burrowing lizard. Invertebrate animals commonly called " Worm s" include annelids (earth Worm s), nematodes (round Worm s), flat Worm s, marine polychaete Worm s (bristle Worm s), marine nemertean Worm ("bootlace Worm s"), and insect larvae such as caterpillars, grubs, and maggots.

Fauna

Fauna is all of the animal life of any particular region or time. The corresponding term for plants is flora.

Zoologists and paleontologists use Fauna to refer to a typical collection of animals found in a specific time or place, e.g. the "Sonoran Desert Fauna" or the "Burgess shale Fauna".

Embryo

An Embryo is a multicellular diploid eukaryote in its earliest stage of development, from the time of first cell division until birth, hatching, or germination. In humans, it is called an Embryo until about eight weeks after fertilization (i.e. ten weeks LMP), and from then it is instead called a fetus. 6 week old excised human Embryo /span>

The development of the Embryo is called Embryo genesis.

Larva

A Larva is a young form of animal with indirect development, going through or undergoing metamorphosis (for example, insects, amphibians, or cnidarians.)

The Larva can look completely different from the adult form, for example, a caterpillar differs from a butterfly. Larva e often have special (Larva l) organs which do not occur in the adult form.

Trace fossils

Trace fossils are geological records of biological activity. Trace fossils may be impressions made on the substrate by an organism: for example, burrows, borings , footprints and feeding marks, and root cavities. The term in its broadest sense also includes the remains of other organic material produced by an organism - for example coprolites or chemical markers - or sedimentological structures produced by biological means - for example, stromatolites.

Animals	Animals are a major group of mostly multicellular, eukaryotic organisms of the kingdom Animalia or Metazoa. Their body plan eventually becomes fixed as they develop, although some undergo a process of metamorphosis later on in their life. Most Animals are motile, meaning they can move spontaneously and independently.
Wiwaxia	Wiwaxia is a genus of soft-bodied, scale-covered animals known from Burgess shale type Lagerstätte dating from the Early to Middle Cambrian. The organisms are mainly known from dispersed sclerites; articulated specimens, where found, range from 3.4 millimetres (0.13 in) to a little over 5 centimetres (2.0 in) in length. The precise taxonomic affinities of the genus are a matter of ongoing debate amongst palaeontologists.
Bilateria	The Bilateria are all animals having a bilateral symmetry, i.e. they have a front and a back end, as well as an upside and downside. Radially symmetrical animals like jellyfish have a topside and downside, but no front and back. The bilateralians are a subregnum (a major group) of animals, including the majority of phyla; the most notable exceptions are the sponges and cnidarians.
Chordates	Chordates are a group of animals that includes the vertebrates, together with several closely related invertebrates. They are united by having, at some time in their life cycle, a notochord, a hollow dorsal nerve cord, pharyngeal slits, an endostyle, and a post-anal tail. The phylum Chordata consists of three subphyla: Urochordata, represented by tunicates; Cephalochordata, represented by lancelets;and Craniata, which includes Vertebrata.
Ctenophora	The Ctenophora , commonly known as comb jellies, are a phylum of animals that live in marine waters worldwide. Their most distinctive feature is the "combs", groups of cilia that they use for swimming, and they are the largest animals that swim by means of cilia -- adults of various species range from a few millimeters to 1.5 meters in size. Like cnidarians, their bodies consist of a mass of jelly with one layer of cells on the outside and another lining the internal cavity.
Deuterostomes	Deuterostomes are a superphylum of animals. They are a subtaxon of the Bilateria branch of the subregnum Eumetazoa, and are opposed to the protostomes. Deuterostomes are distinguished by their embryonic development; in Deuterostomes, the first opening becomes the anus, while in protostomes it becomes the mouth.
Ecdysozoa	The Ecdysozoa are a grouping of protostome animals, including the Arthropoda , Nematoda, and several smaller phyla. They were first defined by Aguinaldo et al. in 1997, based mainly on trees constructed using 18S ribosomal RNA genes.
Endoderm	Endoderm, (sometimes called Entoderm) is one of the germ layers formed during animal embryogenesis. Cells migrating inward along the archenteron from the inner layer of the gastrula, which develops into the Endoderm. The Endoderm consists at first of flattened cells, which subsequently become columnar.

Flatworms

The Flatworms, known in scientific literature as Platyhelminthes are a phylum of relatively simple bilaterian, unsegmented, soft-bodied invertebrate animals. Unlike other bilaterians they have no body cavity, and no specialized circulatory and respiratory organs, which restricts them to flattened shapes that allow oxygen and nutrients to pass through their bodies by diffusion.

In traditional zoology texts Platyhelminthes are divided into Turbellaria, which are mostly non-parasitic animals such as planarians, and three entirely parasitic groups: Cestoda, Trematoda and Monogenea.

Lophotrochozoa

The Lophotrochozoa are a major grouping of protostome animals. The taxon was introduced in 1995 in a paper by Kenneth M Halanych et al. based on molecular data.

Mollusca

Molluscs are animals belonging to the phylum Mollusca. There are around 93,000 recognized extant species, making it the largest marine phylum with about 23% of all named marine organisms. Representatives of the phylum live in a huge range of habitats including marine, freshwater, and terrestrial environments.

Sea star

Sea star s are echinoderms belonging to the class Asteroidea. The names Sea star and "starfish" are sometimes differentiated, with "starfish" used in a broader sense to include the closely related brittle stars, which make up the class Ophiuroidea, as well as excluding Sea star s which do not have five arms (have Many arms), such as the sun stars and cushion stars.

Sea star s exhibit a superficially radial symmetry.

Cleavage

In embryology, Cleavage is the division of cells in the early embryo. The zygotes of many species undergo rapid cell cycles with no significant growth, producing a cluster of cells the same size as the original zygote. The different cells derived from Cleavage are called blastomeres and form a compact mass called the morula.

Clock

CLOCK, or CLOCK is a gene which encodes proteins regulating circadian rhythm. The CLOCK protein seems to affect both the persistence and length of the circadian cycle. CLOCK forms part of a basic-helix-loop-helix transcription factor.

Gastrulation

Gastrulation is a phase early in the development of animal embryos, during which the morphology of the embryo is dramatically restructured by cell migration. Gastrulation varies in different phyla. Gastrulation is followed by organogenesis, when individual organs develop within the newly formed germ layers.

Mesoderm

One of the three germ layers found in the embryos of animals more complex than cnidarians, making them triploblastic. Mesoderm forms in the embryo during gastrulation when some of the cells migrating inward to form the endoderm, produce an additional layer that lies between the endoderm and the ectoderm.

Mesoderm is found in all large, complex animals, and allows the formation of a coelom, which allows more room for independent growth of the body organs.

Molecular clock	The Molecular clock) is a technique in molecular evolution to relate the time that two species diverged to the number of molecular differences measured between the species" DNA sequences or proteins. It is sometimes called a gene clock or evolutionary clock. The notion of the existence of a so-called "Molecular clock" was first attributed to Emile Zuckerkandl and Linus Pauling who, in 1962, noticed that the number of amino acid differences in hemoglobin between different lineages roughly changes with time, as estimated also from fossil evidence.
Protostomes	Protostomia are a clade of animals. Together with the deuterostomes and a few smaller phyla, they make up the Bilateria, mostly comprising animals with bilateral symmetry and three germ layers. The major distinctions between deuterostomes and Protostomes are found in embryonic development.
Echinoderms	Echinoderms are a phylum of marine animals. Echinoderms are found at every ocean depth, from the intertidal zone to the abyssal zone. Aside from the problematic Arkarua, the first definitive members of the phylum appeared near the start of the Cambrian period The phylum contains about 7,000 living species, making it the second-largest grouping of deuterostomes, after the chordates; they are also the largest phylum that has no freshwater or terrestrial representatives.
Oxygen	Oxygen and -γενÎ®ς (-genÄ"s) (producer, literally begetter) is the element with atomic number 8 and represented by the symbol O. It is a member of the chalcogen group on the periodic table, and is a highly reactive nonmetallic period 2 element that readily forms compounds (notably oxides) with almost all other elements. At standard temperature and pressure two atoms of the element bind to form di Oxygen , a colorless, odorless, tasteless diatomic gas with the formula O_2. Oxygen is the third most abundant element in the universe by mass after hydrogen and helium and the most abundant element by mass in the Earth"s crust.
Radiation	In physics, radiation describes any process in which energy emitted by one body travels through a medium or through space, ultimately to be absorbed by another body. Non-physicists often associate the word with ionizing radiation, but it can also refer to electromagnetic radiation (i.e., radio waves, infrared light, visible light, ultraviolet light, and X-rays) which can also be ionizing radiation, to acoustic radiation, or to other more obscure processes. What makes it radiation is that the energy radiates (i.e., it travels outward in straight lines in all directions) from the source.
Plants	Plants are living organisms belonging to the kingdom Plantae. They include familiar organisms such as trees, herbs, bushes, grasses, vines, ferns, mosses, and green algae. About 350,000 species of Plants, defined as seed Plants, bryophytes, ferns and fern allies, are estimated to exist currently.

Charles Robert Darwin	Charles Robert Darwin FRS (12 February 1809 - 19 April 1882) was an English naturalist who realised and presented compelling evidence that all species of life have evolved over time from common ancestors, through the process he called natural selection. The fact that evolution occurs became accepted by the scientific community and much of the general public in his lifetime, while his theory of natural selection came to be widely seen as the primary explanation of the process of evolution in the 1930s, and now forms the basis of modern evolutionary theory. In modified form, Darwin"s scientific discovery is the unifying theory of the life sciences, providing logical explanation for the diversity of life.
Bryozoa	Bryozoans are tiny colonial animals that generally build stony skeletons of calcium carbonate, superficially similar to coral (although some species lack any calcification in the colony and instead have a mucilaginous structure.) Members of the Phylum Bryozoa are known as "moss animals" or "moss animacules" or as "sea mats". They generally prefer warm, tropical waters, but are known to occur worldwide.
Bacteria	The Bacteria are a large group of unicellular microorganisms. Typically a few micrometres in length, Bacteria have a wide range of shapes, ranging from spheres to rods and spirals. Bacteria are ubiquitous in every habitat on Earth, growing in soil, acidic hot springs, radioactive waste, water, and deep in the Earth"s crust, as well as in organic matter and the live bodies of plants and animals.
Horseshoe crab	The Horseshoe crab or Atlantic Horseshoe crab (Limulus polyphemus) is a marine chelicerate arthropod. Despite its name, it is more closely related to spiders, ticks, and scorpions than to crabs. Horseshoe crab s are most commonly found in the Gulf of Mexico and along the northern Atlantic coast of North America.
David	DAVID (the Database for Annotation, Visualization and Integrated Discovery) is a free online bioinformatics resources developed by the Laboratory of Immunopathogenesis and Bioinformatics (LIB.) All tools in the DAVID Bioinformatics Resources aim to provide functional interpretation of large lists of genes derived from genomic studies, e.g. microarray and proteomics studies. DAVID can be found at http://DAVID.niaid.nih.gov or http://DAVID.abcc.ncifcrf.gov The newly updated DAVID Bioinformatics Resources consists of the DAVID Knowledgebase and five integrated, web-based functional annotation tool suites: the DAVID Gene Functional Classification Tool, the DAVID Functional Annotation Tool, the DAVID Gene ID Conversion Tool, the DAVID Gene Name Viewer and the DAVID NIAID Pathogen Genome Browser.
Rat	Rat s are various medium-sized, long-tailed rodents of the superfamily Muroidea. "True Rat s" are members of the genus Rat tus, the most important of which to humans are the black Rat Rat tus Rat tus, and the brown Rat Rat tus norvegicus. Many members of other rodent genera and families are also called Rat s and share many characteristics with true Rat s.

Acid	An acid is traditionally considered any chemical compound that, when dissolved in water, gives a solution with a hydrogen ion activity greater than in pure water, i.e. a pH less than 7.0. That approximates the modern definition of Johannes Nicolaus Brønsted and Martin Lowry, who independently defined an acid as a compound which donates a hydrogen ion (H^+) to another compound (called a base.) Common examples include acetic acid and sulfuric acid (used in car batteries.)
Sulfur	Sulfur or sulphur is the chemical element that has the atomic number 16. It is denoted with the symbol S. It is an abundant, multivalent non-metal. Sulfur in its native form is a yellow crystalline solid.
Sulfur dioxide	Sulfur dioxide is the chemical compound with the formula SO_2. It is produced by volcanoes and in various industrial processes. Since coal and petroleum often contain sulfur compounds, their combustion generates Sulfur dioxide.
Phytoplankton	Phytoplankton are the autotrophic component of the plankton community. The name comes from the Greek words phyton, or "plant", and πλαγκτος , meaning "wanderer" or "drifter". Most Phytoplankton are too small to be individually seen with the unaided eye.
Plankton	Plankton consist of any drifting organisms (animals, plants, archaea seas, or bodies of fresh water. Plankton are defined by their ecological niche rather than their phylogenetic or taxonomic classification. They provide a crucial source of food to more familiar aquatic organisms such as fish.
Fern	A Fern is any one of a group of about 20,000 species of plants classified in the phylum or division Pteridophyta, also known as Filicophyta. The group is also referred to as Polypodiophyta, or Polypodiopsida when treated as a subdivision of tracheophyta (vascular plants.) The term "pteridophyte" has traditionally been used to describe all seedless vascular plants, making it synonymous with " Fern s and Fern allies".
Fern spike	In paleontology, a Fern spike is the occurrence of abundant fern spores in the fossil record, usually immediately (in a geological sense) after an extinction event. The spikes are believed to represent a large, temporary increase in the number of ferns relative to other terrestrial plants after the extinction or thinning of the latter, probably because fern dispersal is more rapid over large geographic areas, since single-celled fern spores are more easily distributed by the wind than are seeds. Fern spike s are most associated with the Cretaceous-Tertiary extinction event, although they have been found at other events such as at the Triassic-Jurassic boundary.
Invertebrate	An Invertebrate is an animal without a vertebral column. The group includes 95% of all animal species -- all animals except those in the Chordate subphylum Vertebrata (fish, reptiles, amphibians, birds, and mammals.) Carolus Linnaeus" Systema Naturae divided these animals into only two groups, the Insecta and the now-obsolete vermes (worms.)

Species	In biology, a Species is: · a taxonomic rank (the basic rank of biological classification) or · a unit at that rank There are many definitions of what kind of unit a Species is (or should be.) A common definition is that of a group of organisms capable of interbreeding and producing fertile offspring, and separated from other such groups with which interbreeding does not (normally) happen. Other definitions may focus on similarity of DNA or morphology. Some Species are further subdivided into sub Species , and here also there is no close agreement on the criteria to be used.
Human	A Human is a member of a species of bipedal primates in the family Hominidae . DNA and fossil evidence indicates that modern Human s originated in east Africa about 200,000 years ago. When compared to other animals and primates, Human s have a highly developed brain, capable of abstract reasoning, language, introspection and problem solving.
Marine invertebrates	The term "Marine invertebrates" is used to describe animals found in a marine environment which are invertebrates: lacking a notochord. In order to protect themselves, they may have evolved a shell or a hard exoskeleton, but this is not always the case. As on land and in the air, invertebrates make up a huge portion of all life in the sea.
Ascertainment bias	In scientific research, Ascertainment bias occurs when false results are produced by non-random sampling and conclusions made about an entire group are based on a distorted or nontypical sample. If this is not accounted for, results can be erroneously attributed to the phenomenon under study rather than to the method of sampling. It is one of the most common reasons that researchers in the medical, social, or biological sciences may discover an association or correlation that does not actually exist.
Population	In biology, a population is the collection of inter-breeding organisms of a particular species; in sociology, a collection of human beings. Individuals within a population share a factor may be reduced by statistical means, but such a generalization may be too vague to imply anything. Demography is used extensively in marketing, which relates to economic units, such as retailers, to potential customers.
AIDS	Acquired immune deficiency syndrome or acquired immunodeficiency syndrome (AIDS) is a disease of the human immune system caused by the human immunodeficiency virus (HIV.) This condition progressively reduces the effectiveness of the immune system and leaves individuals susceptible to opportunistic infections and tumors. HIV is transmitted through direct contact of a mucous membrane or the bloodstream with a bodily fluid containing HIV, such as blood, semen, vaginal fluid, preseminal fluid, and breast milk.

Deforestation	Deforestation is the logging and/or burning of trees in a forested area. There are several reasons deforestation occurs: trees or derived charcoal can be sold as a commodity and used by humans, while cleared land is used as pasture, plantations of commodities and human settlement. The removal of trees without sufficient reforestation has resulted in damage to habitat, biodiversity loss and aridity.

Arabidopsis thaliana	Arabidopsis thaliana , is a small flowering plant native to Europe, Asia, and northwestern Africa. A spring annual with a relatively short life cycle, Arabidopsis is popular as a model organism in plant biology and genetics. Its genome is one of the smallest plant genomes and was the first plant genome to be sequenced.
Hox genes	Hox genes are a group of related genes that specify the anterior-posterior axis and segment identity of metazoan organisms during early embryonic development. These genes are critical for the proper number and placement of embryonic segment structures (such as legs, antennae, and eyes.) Homeotic genes are defined by a DNA sequence known as the homeobox, which is a sequence of 180 nucleotides that code for a protein domain known as the homeodomain.
Indy	Indy, short for I"m not dead yet, is a gene of the model organism, the fruitfly Drosophila melanogaster. Mutant versions of this gene doubles the fruitflies" average lifespan (this is subject to controversy.) Its name originates from a repeated comic line in Monty Python and the Holy Grail .
Mutant	A Mutant is an individual, organism which is a base-pair sequence change within the DNA of a gene or chromosome of an organism resulting in the creation of a new character or trait not found in the wild type. The natural occurrence of genetic mutations is integral to the process of evolution. The study of Mutant s is an integral part of biology, by understanding the effect that a mutation in a gene has it is possible to establish the normal function of that gene.
Mutation	In biology, mutation s are changes to the nucleotide sequence of the genetic material of an organism. mutation s can be caused by copying errors in the genetic material during cell division, by exposure to ultraviolet or ionizing radiation, chemical mutagens, or viruses, or can be induced by the organism itself, by cellular processes such as hyper mutation . In multicellular organisms with dedicated reproductive cells, mutation s can be subdivided into germ line mutation s, which can be passed on to descendants through the reproductive cells, and somatic mutation s, which involve cells outside the dedicated reproductive group and which are not usually transmitted to descendants.
History of the Polymerase Chain Reaction	The history of the Polymerase Chain Reaction has variously been described as a classic "Eureka!" moment, or as an example of cooperative teamwork between disparate researchers. A list of some of the events before, during, and after its development: · On April 25, 1953 James D. Watson and Francis Crick publish "a radically different structure" for DNA, thereby founding the field of Molecular Genetics. Their structure involves two strands of complementary base-paired DNA, running in opposite directions as a double helix. They conclude their report saying that "It has not escaped our notice that the specific pairing we have postulated immediately suggests a possible copying mechanism for the genetic material". They are awarded the Nobel Prize in 1962.

· Starting in the mid 1950s, Arthur Kornberg begins to study the mechanism of DNA replication. By 1957 he has identified the first DNA polymerase. The enzyme is surprisingly limited, creating DNA in just one direction and requiring an existing primer to initiate copying of the template strand.

Polymerase

A Polymerase (EC 2.7.7.6/7/19/48/49) is an enzyme whose central function is associated with polymers of nucleic acids such as RNA and DNA. The primary function of a Polymerase is the polymerization of new DNA or RNA against an existing DNA or RNA template in the processes of replication and transcription. In association with a cluster of other enzymes and proteins, they take nucleotides from solvent, and catalyse the synthesis of a polynucleotide sequence against a nucleotide template strand using base-pairing interactions.

It is an accident of history that the enzymes responsible for the catalytic production of other biopolymers are not also referred to as Polymerase s.

Polymerase chain reaction

In molecular biology, the Polymerase chain reaction is a technique to amplify a single or few copies of a piece of DNA across several orders of magnitude, generating millions or more copies of a particular DNA sequence. The method relies on thermal cycling, consisting of cycles of repeated heating and cooling of the reaction for DNA melting and enzymatic replication of the DNA. Primers (short DNA fragments) containing sequences complementary to the target region along with a DNA polymerase (after which the method is named) are key components to enable selective and repeated amplification. As Polymerase chain reaction progresses, the DNA generated is itself used as a template for replication, setting in motion a chain reaction in which the DNA template is exponentially amplified.

Gene

A Gene is the basic unit of heredity in a living organism. All living things depend on Gene s. Gene s hold the information to build and maintain their cells and pass Gene tic traits to offspring.

Homeotic genes

Homeotic genes are genes that determine which parts of the body form what body parts. One example is the Hox genes which are important for segmentation, another example are the MADS-box-containing genes in the ABC model of flower development.

MADS-box

MADS box is a conserved sequence motif found in a family of transcription factors, the MADS-box protein family. The length of the MADS-box reported by various researchers varies somewhat, but typical lengths are in the range of 168 to 180 base pairs.

The name is an acronym that refers to the genes in which the sequence element was first identified:

· MCM1 from the budding yeast, Saccharomyces cerevisiae,
· AGAMOUS from Arabidopsis thaliana,
· DEFICIENS from the snapdragon Antirrhinum majus,
· SRF from Homo sapiens.

	The element encodes the MADS-domain that have DNA-binding properties. In plants, MADS-box genes have undergone a substantial radiation and include among others floral homeotic genes that participate in the determination of floral organ identity according to The ABC Model of Flower Development.
Animals	Animals are a major group of mostly multicellular, eukaryotic organisms of the kingdom Animalia or Metazoa. Their body plan eventually becomes fixed as they develop, although some undergo a process of metamorphosis later on in their life. Most Animals are motile, meaning they can move spontaneously and independently.
Body plan	A Body plan is essentially the blueprint for the way the body of an organism is laid out. An organism"s symmetry, its number of body segments and number of limbs are all aspects of its Body plan One of the key issues of developmental biology is the evolution of Body plan s as different as those of a starfish, a fern, or a mammal, from a common biological heritage, and in particular how radical changes in Body plan s have occurred over geological time.
Chromosome	A Chromosome is an organized structure of DNA and protein that is found in cells. It is a single piece of coiled DNA containing many genes, regulatory elements and other nucleotide sequences. Chromosome s also contain DNA-bound proteins, which serve to package the DNA and control its functions.
Inversion	An inversion is a chromosome rearrangement in which a segment of a chromosome is reversed end to end. An inversion occurs when a single chromosome undergoes breakage and rearrangement within itself. Inversions are of two types: paracentric and pericentric.
Allele	An Allele) is one of a series of different forms of a gene. The word is a short from of allelomorph , which was used in the early days of genetics to describe variant forms of a gene detected as different phenotypes. Allele s are now understood to be alternative DNA sequences at the same physical gene locus, which may or may not result in different phenotypic traits.
Allele frequency	Allele frequency is the proportion of all variants (alleles) that is made up of a particular gene variant. In other words, it is the number of copies of a particular allele divided by the number of copies of all alleles at the genetic place (locus) in a population. It can be expressed for example as a percentage.
Antennapedia	Antennapedia is a hox gene first discovered in Drosophila which controls the placement of legs. A loss-of-function mutation in the regulatory region of this gene can result in the development of the second leg pair into ectopic antennae. By contrast gain-of-function alleles convert antenna into ectopic legs.

Bithorax complex	The Bithorax complex is a group of homeotic genes in Drosophila melanogaster which are believed to control the differentiation of the abdominal and posterior thoracic segments, located on chromosome III. When these genes are mutated, the third thoracic segment becomes a repeat of the second thoracic segment, creating what is essentially a second thorax. This can result in a second pair of wings, a second stomach, and duplicated thoracic features in varying degrees .
Drosophila	Drosophila has long been a favorite model system for geneticists and developmental biologists studying embryogenesis. The small size, short generation time, and large brood size makes it ideal for genetic studies. Transparent embryos facilitate developmental studies.
Homeobox	A Homeobox is a DNA sequence found within genes that are involved in the regulation of patterns of development (morphogenesis) in animals, fungi and plants. They were discovered independently in 1983 by Walter Jakob Gehring and his colleagues at the University of Basel, Switzerland, and Matthew P. Scott and Amy Weiner, who were then working with Thomas Kaufman at Indiana University in Bloomington. A Homeobox is about 180 base pairs long.
Gene product	A Gene product is the biochemical material, either RNA or protein, resulting from expression of a gene. A measurement of the amount of Gene product is sometimes used to infer how active a gene is. Abnormal amounts of Gene product can be correlated with disease-causing alleles, such as the overactivity of oncogenes which can cause cancer.
Nematode	The "roundworms" or " Nematode s" (phylum Nematoda) are the most diverse phylum of pseudocoelomates, and one of the most diverse of all animals. Nematode species are very difficult to distinguish; over 80,000 have been described, of which over 15,000 are parasitic. It has been estimated that the total number of described and undescribed roundworms might be more than 500,000.
Worm	The term Worm is used to describe many different distantly-related animals which have a long cylindrical body and no legs. Most animals called " Worm s" are invertebrates, but the term is also used for the amphibian caecilians and the slow Worm Anguis, a legless burrowing lizard. Invertebrate animals commonly called " Worm s" include annelids (earth Worm s), nematodes (round Worm s), flat Worm s, marine polychaete Worm s (bristle Worm s), marine nemertean Worm ("bootlace Worm s"), and insect larvae such as caterpillars, grubs, and maggots.
Echinoderms	Echinoderms are a phylum of marine animals. Echinoderms are found at every ocean depth, from the intertidal zone to the abyssal zone. Aside from the problematic Arkarua, the first definitive members of the phylum appeared near the start of the Cambrian period The phylum contains about 7,000 living species, making it the second-largest grouping of deuterostomes, after the chordates; they are also the largest phylum that has no freshwater or terrestrial representatives.

Vertebrates	Vertebrates are members of the subphylum Vertebrata, chordates with backbones or spinal columns. The grouping sometimes includes the hagfish, which have no vertebrae, but are genetically quite closely related to lampreys, which do have vertebrae. For this reason, the subphylum is sometimes referred to as "Craniata", as all members do possess a cranium.
Blytt-Sernander	The Blytt-Sernander classification is a series of north European climatic periods or phases based on the study of Danish peat bogs by Axel Blytt (1876) and Rutger Sernander (1908.) The classification was incorporated into a sequence of pollen zones later defined by Lennart von Post, one of the founders of palynology. Layers in peat were first noticed by Heinrich Dau in 1829.
Crustaceans	Crustaceans are a very large group of arthropods, comprising almost 52,000 described species , and are usually treated as a subphylum . They include various familiar animals, such as crabs, lobsters, crayfish, shrimp, krill and barnacles. The majority of them are aquatic, living in either marine or fresh water environments, but a few groups have adapted to life on land, such as terrestrial crabs, terrestrial hermit crabs and woodlice.
Eurypterid	Eurypterid s (sea scorpions) are an extinct group of arthropods related to arachnids, which include the largest known arthropods that ever lived. They are members of the extinct class Eurypterid a (Chelicerata.) The word Eurypterid comes from the Greek word eury meaning "broad" or "wide" and the Greek word pteron meaning "wing" .
Sea spiders	Sea spiders are marine arthropods of class Pycnogonida. They are cosmopolitan, found especially in the Mediterranean and Caribbean Seas and the Arctic and Antarctic Oceans. There are over 1300 known species, ranging in size from 1 to 10 millimetres, to over 90 centimetres in some deep water species.
Velvet worms	The velvet worms form a phylum within the Ecdysozoa and can be simply described as "worms with legs". Most common in the Southern Hemisphere, they prey on smaller animals such as insects, which they catch by squirting a sticky slime. In modern zoology they are particularly renowned for their curious mating behaviour and for bearing live young.
Heritability	In genetics, heritability is the proportion of phenotypic variation in a population that is attributable to genetic variation among individuals. Variation among individuals may be due to genetic and/or environmental factors. heritability analyses estimate the relative contributions of differences in genetic and non-genetic factors to the total phenotypic variance in a population.

Insects

Insects are arthropods, having a hard exoskeleton, a three-part body (head, thorax, and abdomen), three pairs of jointed legs, compound eyes, and two antennae. They are the most diverse group of animals on the planet and include approximately 2,200 species of praying mantis, 5,000 dragonfly, 20,000 grasshopper, 82,000 true bug, 120,000 fly, 110,000 bee, wasp, ant and sawfly, 170,000 butterfly and moth, and 360,000 beetle species described to date. The number of extant species is estimated at between six and ten million, with over a million species already described.

Gene expression

Gene expression is the process by which information from a gene is used in the synthesis of a functional gene product. These products are often proteins, but in non-protein coding genes such as rRNA genes or tRNA genes, the product is a functional RNA.

Several steps in the Gene expression process may be modulated, including the transcription, RNA splicing, translation, and post-translational modification of a protein. Gene regulation gives the cell control over structure and function, and is the basis for cellular differentiation, morphogenesis and the versatility and adaptability of any organism.

Fauna

Fauna is all of the animal life of any particular region or time. The corresponding term for plants is flora.

Zoologists and paleontologists use Fauna to refer to a typical collection of animals found in a specific time or place, e.g. the "Sonoran Desert Fauna" or the "Burgess shale Fauna".

Genetic

Genetics is the study of how living things receive common traits from previous generations. These traits are described by the Genetic information carried by a molecule called DNA. The instructions for constructing and operating an organism are contained in the organism"s DNA. Every living thing on earth has DNA in its cells. Genes are the hereditary components of DNA that occupy spots on chromosomes and determine characteristics in an organism.

Fin

A Fin is a surface used to produce lift and thrust or to steer while traveling in water, air, or other fluid media. The first use of the word was for the limbs of fish, but has been extended to include other animal limbs and man-made devices.

The foremost use of Fin s is to ensure the directional stability of an object moving through a fluid such as water or air and may be seen in the use of fletching on arrows and Fin s at the rear of some missiles, rockets, self-propelled torpedoes, and kinetic energy penetrators.

Fish

FISH (fluorescence in situ hybridization) is a cytogenetic technique used to detect and localize the presence or absence of specific DNA sequences on chromosomes. FISH uses fluorescent probes that bind to only those parts of the chromosome with which they show a high degree of sequence similarity. Fluorescence microscopy can be used to find out where the fluorescent probe bound to the chromosomes.

Apical ectodermal ridge

The Apical ectodermal ridge is a critical component in vertebrate limb development. The Apical ectodermal ridge is an ectodermal structure overlying and inducing the developing limb bud of the vertebrate embryo, and will eventually give rise to the skin covering the limb. Initial formation of the Apical ectodermal ridge is induced by secretion of the fibroblast growth factor FGF-10 from the somatic layer of the lateral plate mesoderm.

Fibroblast

A Fibroblast is a type of cell that synthesizes the extracellular matrix and collagen, the structural framework (stroma) for animal tissues, and play a critical role in wound healing. They are the most common cells of connective tissue in animals.

Fibroblast s and fibrocytes are two states of the same cells, the former being the activated state, the latter the less active state, concerned with maintenance.

Fibroblast growth factor

Fibroblast growth factor s are a family of growth factors involved in angiogenesis, wound healing, and embryonic development. The Fibroblast growth factor s are heparin-binding proteins and interactions with cell-surface associated heparan sulfate proteoglycans have been shown to be essential for Fibroblast growth factor signal transduction. Fibroblast growth factor s are key players in the processes of proliferation and differentiation of wide variety of cells and tissues.

Wnt

The Wnt signaling pathway describes a complex network of proteins most well known for their roles in embryogenesis and cancer, but also involved in normal physiological processes in adult animals.

The name Wnt was coined as a combination of Wg (wingless) and Int and can be pronounced as "wint". The wingless gene had originally been identified as a recessive mutation affecting wing and haltere development in Drosophila melanogaster.

Limb development

Limb development in tetrapods -- animals with four limbs -- is an area of active research in developmental biology. Limb formation begins in the limb field, as a limb "bud." Fibroblast growth factor (FGF) induces formation of an organizer, called the apical ectodermal ridge (AER), which guides further development and controls cell death. Apoptosis -- programmed cell death -- is necessary to eliminate webbing between digits.

Appendage

An Appendage in the broadest sense is an additional or subsidiary part existing on something which can generally still function if the Appendage has never existed or is later provided or grown or natural prolongation, that protrudes from an organism"s body, such as a vertebrate"s limbs.

In invertebrate biology, "Appendage" is a general term that covers any of the homologous body parts that may extend from a body segment.

Form

In botanical nomenclature, a Form is a low-level taxonomic rank below that of variety; it is an infraspecific taxon. Its name consists of three parts: a genus name, a specific epithet, and an infraspecific epithet. The abbreviation "f." or the full "forma" should be put before the infraspecific epithet to indicate the rank.

DNA

Deoxyribonucleic acid (DNA) is a nucleic acid that contains the genetic instructions used in the development and functioning of all known living organisms and some viruses. The main role of DNA molecules is the long-term storage of information. DNA is often compared to a set of blueprints or a recipe, or a code, since it contains the instructions needed to construct other components of cells, such as proteins and RNA molecules.

Engrailed

Engrailed is a gene involved in many aspects of multicellular development. First known for its role in arthropod embryological development, working in consort with the Hox genes, Engrailed has been found to be important in other areas of development. It has also been found to direct development in other taxa, for example: humans.

Peripatopsis

Peripatopsis is a genus of velvet worm in the Peripatopsidae family. It includes the following species:

· White cave velvet worm (Peripatopsis alba)
· Knysna velvet worm (Peripatopsis clavigera)
· Lion"s Hill velvet worm (Peripatopsis leonina) .

Carpel

Carpel s are the building blocks of the gynoecium. If a gynoecium has a single carpel it is called monocarpous. If a gynoecium has multiple, distinct (free, unfused) carpel s, it is apocarpous.

Cuticle

A Cuticle is any of a variety of tough but flexible, non-mineral outer coverings of an organism that provide protection. Cuticle s are non-homologous, differing in their origin, structure and chemical composition. Eponychium is the anatomical term for the human Cuticle /span>

In human anatomy, Cuticle refers to several structures.

Fern

A Fern is any one of a group of about 20,000 species of plants classified in the phylum or division Pteridophyta, also known as Filicophyta. The group is also referred to as Polypodiophyta, or Polypodiopsida when treated as a subdivision of tracheophyta (vascular plants.) The term "pteridophyte" has traditionally been used to describe all seedless vascular plants, making it synonymous with " Fern s and Fern allies".

Flower

A Flower sometimes known as a bloom or blossom, is the reproductive structure found in Flower ing plants The biological function of a Flower is to mediate the union of male sperm with female ovum in order to produce seeds. The process begins with pollination, is followed by fertilization, leading to the formation and dispersal of the seeds.

Gymnosperm	Gymnosperm (Gymnosperm ae) is a group of spermatophyte seed-bearing plants with ovules on scales, which are usually arranged in cone-like structures. The other major group of seed-bearing plants, the angiosperms , have ovules enclosed in a carpel, a sporophyll with fused margins. A carpel consists of a stigma, style and ovary.
Petal	A Petal is one member or part of the corolla of a flower. The corolla is the name for all of the Petal s of a flower; the inner perianth whorl, term used when this is not the same in appearance as the outermost whorl (the calyx) and is used to attract pollinators based on its bright color. It is the inner part of the perianth that comprises the sterile parts of a flower and consists of inner and outer tepals.
Plants	Plants are living organisms belonging to the kingdom Plantae. They include familiar organisms such as trees, herbs, bushes, grasses, vines, ferns, mosses, and green algae. About 350,000 species of Plants, defined as seed Plants, bryophytes, ferns and fern allies, are estimated to exist currently.
Pollen	Pollen is a fine to coarse powder consisting of microgametophytes (Pollen grains), which produce the male gametes (sperm cells) of seed plants. A hard coat covering the Pollen grain protects the sperm cells during the process of their movement between the stamens of the flower to the pistil of the next flower. Individual Pollen grains are small enough to require magnification to see detail.
Seed	A Seed, referred to as a kernel in some plants, is a small embryonic plant enclosed in a covering called the Seed coat, usually with some stored food. It is the product of the ripened ovule of gymnosperm and angiosperm plants which occurs after fertilization and some growth within the mother plant. The formation of the Seed completes the process of reproduction in Seed plants (started with the development of flowers and pollination), with the embryo developed from the zygote and the Seed coat from the integuments of the ovule.
Sepal	A Sepal is a part of the flower of angiosperms . Sepal s in most flowers are green and lie under the more conspicuous petals. As a collective unit the Sepal s form a calyx, whereas the collection of petals is called the corolla.
Stamen	The Stamen ">warp") is the male organ of a flower. Each Stamen generally has a stalk called the filament , and, on top of the filament, an anther , and pollen sacs, called microsporangia. The development of the microsporangia and the contained haploid gametophytes, (called pollen-grains) is closely comparable with that of the microsporangia in gymnosperms or heterosporous ferns.
Vascular tissue	Vascular tissue is a complex conducting tissue, formed of more than one cell type, found in vascular plants. The primary components of Vascular tissue are the xylem and phloem. These two tissues transport fluid and nutrients internally.

Cell	The Cell is the structural and functional unit of all known living organisms. It is the smallest unit of an organism that is classified as living, and is often called the building block of life. Some organisms, such as most bacteria, are unicellular (consist of a single Cell.)
AGAMOUS	Agamous is a gene and transcription factor for Arabidopsis Thaliana.

Human	A Human is a member of a species of bipedal primates in the family Hominidae . DNA and fossil evidence indicates that modern Human s originated in east Africa about 200,000 years ago. When compared to other animals and primates, Human s have a highly developed brain, capable of abstract reasoning, language, introspection and problem solving.
Species	In biology, a Species is: · a taxonomic rank (the basic rank of biological classification) or · a unit at that rank There are many definitions of what kind of unit a Species is (or should be.) A common definition is that of a group of organisms capable of interbreeding and producing fertile offspring, and separated from other such groups with which interbreeding does not (normally) happen. Other definitions may focus on similarity of DNA or morphology. Some Species are further subdivided into sub Species , and here also there is no close agreement on the criteria to be used.
Chimpanzee	Chimpanzee sometimes colloquially chimp, is the common name for the two extant species of ape in the genus Pan where the Congo River forms the boundary between the native habitat of the two species: · Common Chimpanzee Pan troglodytes: the better known Chimpanzee lives primarily in West and Central Africa. · Bonobo, Pan paniscus: also known as the "Pygmy Chimpanzee , this species is found in the forests of the Democratic Republic of the Congo. Chimpanzee s are members of the Hominidae family, along with gorillas, humans, and orangutans. Chimpanzee are thought to have split from human evolution about 6 million years ago and thus the two Chimpanzee species are the closest living relatives to humans, all being members of the Hominini tribe (along with extinct species of Hominina subtribe.) Chimpanzee s are the only known members of the Panina subtribe. The two Pan species split only about one million years ago.
Orangutans	The Orangutans are a species of great apes. Known for their intelligence, they live in trees and are the largest living arboreal animal. They have longer arms than other great apes, and their hair is typically reddish-brown, instead of the brown or black hair typical of other great apes.
RNA	Ribonucleic acid (RNA) is a biologically important type of molecule that consists of a long chain of nucleotide units. Each nucleotide consists of a nitrogenous base, a ribose sugar, and a phosphate. RNA is very similar to DNA, but differs in a few important structural details: in the cell, RNA is usually single-stranded, while DNA is usually double-stranded; RNA nucleotides contain ribose while DNA contains deoxyribose (a type of ribose that lacks one oxygen atom); and RNA has the base uracil rather than thymine that is present in DNA. RNA is transcribed from DNA by enzymes called RNA polymerases and is generally further processed by other enzymes.

Albumin	Albumin refers generally to any protein with water solubility, which is moderately soluble in concentrated salt solutions, and experiences heat coagulation (protein denaturation.) Substances containing Albumin such as egg white, are called Albumin oids. It has a molecular weight of about 65kD (65,000 Atomic mass units) and consists of 584 amino acids and contains no carbohydrate.
Blytt-Sernander	The Blytt-Sernander classification is a series of north European climatic periods or phases based on the study of Danish peat bogs by Axel Blytt (1876) and Rutger Sernander (1908.) The classification was incorporated into a sequence of pollen zones later defined by Lennart von Post, one of the founders of palynology. Layers in peat were first noticed by Heinrich Dau in 1829.
Cytochromes	Cytochromes are, in general, membrane-bound hemoproteins that contain heme groups and carry out electron transport. They are found either as monomeric proteins (e.g., cytochrome c) or as subunits of bigger enzymatic complexes that catalyze redox reactions. They are found in the mitochondrial inner membrane and endoplasmic reticulum of eukaryotes, in the chloroplasts of plants, in photosynthetic microorganisms, and in bacteria.
DNA	Deoxyribonucleic acid (DNA) is a nucleic acid that contains the genetic instructions used in the development and functioning of all known living organisms and some viruses. The main role of DNA molecules is the long-term storage of information. DNA is often compared to a set of blueprints or a recipe, or a code, since it contains the instructions needed to construct other components of cells, such as proteins and RNA molecules.
Mitochondrial DNA	Mitochondrial DNA is the DNA located in organelles called mitochondria. Most other DNA present in eukaryotic organisms is found in the cell nucleus. Mitochondrial DNA was discovered by Margit M. K. Nass and Sylvan Nass by electron microscopy as DNAase-sensitive thread inside mitochondria, and by Ellen Haslbrunner, Hans Tuppy and Gottfried Schatz by biochemical assays on highly purified mitochondrial fractions.
Gene	A Gene is the basic unit of heredity in a living organism. All living things depend on Gene s. Gene s hold the information to build and maintain their cells and pass Gene tic traits to offspring.
Endogenous retroviruses	Endogenous retroviruses are retroviruses derived from ancient viral infections of germ cells in humans, mammals and other vertebrates; as such their proviruses are passed on to the next generation and now remain in the genome. Retroviruses are viruses that reverse-transcribe their RNA into DNA for integration into the host"s genome. Most retroviruses (such as HIV-1) infect somatic cells, but some can also infect germline cells (cells that make eggs and sperm) and once they have done so and have been transmitted to the next generation, they are termed endogenous.

Genome	In classical genetics, the Genome of a diploid organism including eukarya refers to a full set of chromosomes or genes in a gamete; thereby, a regular somatic cell contains two full sets of Genome s. In haploid organisms, including bacteria, archaea, viruses, and mitochondria, a cell contains only a single set of the Genome usually in a single circular or contiguous linear DNA (or RNA for retroviruses.) In modern molecular biology the Genome of an organism is its hereditary information encoded in DNA (or, for retroviruses, RNA.)
Human genome	The Human genome is the genome of Homo sapiens, which is stored on 23 chromosome pairs. Twenty-two of these are autosomal chromosome pairs, while the remaining pair is sex-determining. The haploid Human genome occupies a total of just over 3 billion DNA base pairs.
Animals	Animals are a major group of mostly multicellular, eukaryotic organisms of the kingdom Animalia or Metazoa. Their body plan eventually becomes fixed as they develop, although some undergo a process of metamorphosis later on in their life. Most Animals are motile, meaning they can move spontaneously and independently.
David	DAVID (the Database for Annotation, Visualization and Integrated Discovery) is a free online bioinformatics resources developed by the Laboratory of Immunopathogenesis and Bioinformatics (LIB.) All tools in the DAVID Bioinformatics Resources aim to provide functional interpretation of large lists of genes derived from genomic studies, e.g. microarray and proteomics studies. DAVID can be found at http://DAVID.niaid.nih.gov or http://DAVID.abcc.ncifcrf.gov The newly updated DAVID Bioinformatics Resources consists of the DAVID Knowledgebase and five integrated, web-based functional annotation tool suites: the DAVID Gene Functional Classification Tool, the DAVID Functional Annotation Tool, the DAVID Gene ID Conversion Tool, the DAVID Gene Name Viewer and the DAVID NIAID Pathogen Genome Browser.
Acid	An acid is traditionally considered any chemical compound that, when dissolved in water, gives a solution with a hydrogen ion activity greater than in pure water, i.e. a pH less than 7.0. That approximates the modern definition of Johannes Nicolaus Brønsted and Martin Lowry, who independently defined an acid as a compound which donates a hydrogen ion (H^+) to another compound (called a base.) Common examples include acetic acid and sulfuric acid (used in car batteries.)
Genetic	Genetics is the study of how living things receive common traits from previous generations. These traits are described by the Genetic information carried by a molecule called DNA. The instructions for constructing and operating an organism are contained in the organism"s DNA. Every living thing on earth has DNA in its cells. Genes are the hereditary components of DNA that occupy spots on chromosomes and determine characteristics in an organism.
Nick	A Nick is a discontinuity in a double stranded DNA molecule where there is no phosphodiester bond between adjacent nucleotides of one strand typically through damage or enzyme action.

Platyops	Platyops sterreri is a species of crustacean in family Mysidae, endemic to Bermuda, the only species in the genus Platyops .
Size	Size has been one of the most interesting aspects of cephalopod science to the general public Extinct taxa are also included.
White	White was the first sex-linked mutation ever discovered in Drosophila melanogaster. In 1910 Thomas Hunt Morgan, (or, legend has it, his wife) collected a single male White-eyed mutant from a population of Drosophila melanogaster fruit flies, which usually have bright red eyes. Upon breeding this male with wild-type female flies he found that the offspring did not conform to the expectations of Mendelian inheritance.
Assimilation	Biological Assimilation, or bioassimilation, involves one of two different processes to supply animal cells with nutrients. The first is the process of absorbing vitamins, minerals, and other chemicals from food within the gastrointestinal tract. In humans this is done with a chemical breakdown (enzymes and acids) and physical breakdown (oral mastication and stomach churning.)
Genetic diversity	Genetic diversity is a level of biodiversity that refers to the total number of genetic characteristics in the genetic makeup of a species. It is distinguished from genetic variability, which describes the tendency of genetic characteristics to vary. The academic field of population genetics includes several hypotheses and theories regarding genetic diversity.
Noah	Noah was the name of species of ox called gaur, cloned and gestated in the womb of a cow named Bessie. Gaur is a vulnerable species according to the IUCN. The baby bull gaur was delivered on January 8, 2001, but died within just 48 hours of a common dysentery on January 10, 2001. Noah"s condition was monitored by Dr. Jonathan Hill and his teammates in Iowa.
Population	In biology, a population is the collection of inter-breeding organisms of a particular species; in sociology, a collection of human beings. Individuals within a population share a factor may be reduced by statistical means, but such a generalization may be too vague to imply anything. Demography is used extensively in marketing, which relates to economic units, such as retailers, to potential customers.
Mate	Mate is a traditional South American infusion. It is prepared from steeping dried leaves of yerba Mate in hot water. It is the national drink in Argentina, Paraguay, and Uruguay, and drinking it is a common social practice in parts of Brazil, Chile, eastern Bolivia, Lebanon, Syria, Argentina and Turkey.

Mitochondrial Eve

Mitochondrial Eve is the name given by researchers to the woman who is defined as the matrilineal most recent common ancestor (MRCA) for all currently living humans. Passed down from mother to offspring, her mitochondrial DNA (mtDNA) is now found in all living humans: every mtDNA in every living person is derived from hers. Mitochondrial Eve is the female counterpart of Y-chromosomal Adam, the patrilineal most recent common ancestor, although they lived at different times.

Microsatellite

Microsatellite s are polymorphic loci present in nuclear and organellar DNA that consist of repeating units of 1-6 base pairs in length. They are typically neutral, co-dominant and are used as molecular markers which have wide-ranging applications in the field of genetics, including kinship and population studies. Microsatellite s can also be used to study gene dosage (looking for duplications or deletions of a particular genetic region.)

Polymorphism

Polymorphism in biophysics is the aspect of the behaviour of lipids that influences their long-range order, i.e. how they aggregate. This can be in the form of spheres of lipid molecules (micelles), pairs of layers that face one another (lamellar phase, observed in biological system as a lipid bilayer), a tubular arrangement (hexagonal), or various cubic phases (Fd3m, Im3m, Ia3m, Pn3m, and Pm3m being those discovered so far.) More complicated aggregations have also been observed, rhombohedral, tetragonal and orthorhombic phases have been observed.

Alu sequence

An Alu sequence is a short stretch of DNA originally characterized by the action of the Alu restriction endonuclease. Alu sequence s of different kinds occur in large numbers in primate genomes. In fact, Alu sequence s are the most abundant mobile elements in the human genome.

Genetic drift

Genetic drift or allelic drift is the change in the relative frequency with which a gene variant (allele) occurs in a population that results from the fact that alleles in offspring are a random sample of those in the parents, and because of the role of chance in determining whether a given individual survives and reproduces. A population"s allele frequency is the fraction of the gene copies that share a particular form.

Genetic drift is one of several evolutionary processes which lead to changes in allele frequencies over time.

Trait

A trait is a distinct variant of a phenotypic character of an organism that may be inherited, environmentally determined or somewhere in between. For example, eye color is a character or abstraction of an attribute, while blue, brown and hazel are traits.

A trait may be any single feature or quantifiable measurement of an organism.

Anatomy	The anatomy of spiders is in some aspects similar, but also different from that of other arthropods. The following characteristics are common to all spiders: A body with two segments, eight legs, spinnerets, no chewing parts, no wings, and the presence of chelicerae, which spiders use to hold prey, and in most cases, inject venom. Spiders have non-compound eyes, with most species having eight; the spiders known as Haplogynae may have six or fewer, and certain cave-dwelling spiders may have none at all.
Cortex	In botany, the Cortex is the outer of the stem or root of a plant, bounded on the outside by the epidermis and on the inside by the endodermis. It is composed mostly of undifferentiated cells, usually large thin-walled parenchyma cells of the ground tissue system. The outer cortical cells often acquire irregularly thickened cell walls, and are called collenchyma cells.
Chicken anaemia virus	Chicken anaemia virus is a virus that affects poultry. Chicken anaemia virus causes anaemia, bone marrow atrophy, and severe immunosuppression. Clinical signs of infection of Chicken anaemia virus is predominantly found in young chicks due to maternal antibodies present in most adult chickens.

LaVergne, TN USA
03 September 2010
195792LV00001B/24/P